WORLDS BEFORE OUR OWN

WORLDS BEFORE OUR OWN

BRAD STEIGER

Published by

BERKLEY PUBLISHING CORPORATION

Distributed by

G.P. PUTNAM'S SONS, New York

All rights reserved. This book, or parts thereof, must not be
reproduced in any form without permission. Published simultaneously
in Canada by Longman Canada Limited, Toronto.

SBN: 399-12215-X

Library of Congress Cataloging in Publication Data

Steiger, Brad.
  Worlds before our own.

  Bibliography
  1.   Civilization, Ancient—Extraterrestrial influences.
2.   Man, Prehistoric.   3.   Lost continents.   I.   Title.
CB156.S73   1978   001.9′42   78-18232

PRINTED IN THE UNITED STATES OF AMERICA

CONTENTS

CHAPTER ONE

Worlds Before Our Own

It is rather amazing that such sophisticated people, as we judge ourselves to be, do not even know who we are. And it becomes rather dismaying to discover that there is a great deal of suppressed, ignored, and misplaced prehistorical cultural evidence that would alter the established interpretations of human origins and provide us with a much clearer definition of what it means to be man.

Archaeologists, anthropologists, and various academicians who play the "Origins of Man" game, reluctantly and only occasionally acknowledge instances where unique skeletal and cultural evidence from the prehistoric record suddenly appear long before they should—and in places where they should not. These irritating artifacts destroy the orderly evolutionary line that academia has for so long presented to the public. Consequently, such data has been largely left buried in site reports, forgotten storage rooms, and dusty archives.

Although the Leakeys, the eminent family of anthropologists, have offered dramatic new evidence that the "Homo" lineage goes back at least three million years, the academic consensus holds that an ancestor of modern man evolved about one million years ago. *Homo sapiens*, "thinking man," our own species, became the dominant planetary life form on a worldwide basis about 40,000 years ago.

It is difficult enough to explain the sudden appearance or influx of *Homo sapiens* at that time, but it is an even more complex question to ponder why Neanderthal and Cro-Magnon man correspondingly disappeared. And just when Dr. Richard Leakey is adding to a growing body of evidence that mankind developed in Africa, a Hungarian excavation surrenders a *Homo sapiens* skull fragment in a context more than 600,000 years out of alignment with the accepted calendar of man's migrations across the planet.

The Creationist may present an extreme point of view when he maintains that the world is roughly only 6,000 years old and that man himself is only a few days younger, but what happens to evolution when there are such sites as the one in Australia, which yielded *Homo sapiens* (modern man), *Homo erectus* (our million-year-old ancestor), and Neanderthal (our Stone Age cousin) in what appears to be a contemporaneous environment? Then there is the Tabun site where *Homo sapiens* fragments were found in strata below (which means older than) classic Neanderthal bones. This is but one of several digs that has produced evidence of modern man before what is accepted as one of his predecessors.

Somewhere in what would appear to be a biological and cultural free-for-all there must lie the answer to that most important question: *Who are we?*

But just as we are trying our best to fit skeletal fragments together in a manner that will be found acceptable to what we believe we know about our origins, "pre-Adamite" fossilized footprints are being found, which, if they are what they appear to be, will make total shambles of our accepted evolutionary calendar.

In Pershing County, Nevada, a shoe print was found in

Triassic limestone, strata indicative of 400 million years, in which the fossilized evidence clearly revealed finely wrought double-stitching in the seams.

Early in 1975 Dr. Stanley Rhine of the University of New Mexico announced his discovery of humanlike footprints in strata indicative of 40 million years old.

A few months before, a similar find was made in Kenton, Oklahoma. At almost the same time, such a discovery was revealed in north-central Wisconsin.

At Glen Rose, Texas, a 16-inch hominid footprint was found next to dinosaur tracks in contemporaneous strata, thereby suggesting that man may have outlived the giant reptiles by 80 million years.

If man lived at the same time as the dinosaurs, he might have been a king-sized fellow to better confront the gigantic reptiles. Skeletal remains of surprisingly large human beings have been discovered all over the Americas, from Minnesota to Nicaragua. The skeletons average in size from seven to over eight feet tall—and none of them are carved "Cardiff Giant" hoaxes.

In Death Valley, Utah, there is ample fossil and skeletal evidence to indicate that the desolate area was once a tropical Garden of Eden where a race of giants lived and fed themselves with palatable foods taken from the local lakes and forests.

To speak of a race of prehistoric giants in what is now the desert sands of Death Valley is simultaneously to refute the doctrine which decrees that man is a relative newcomer to the North and South American continents. While on the one hand new radiocarbon dates demonstrate that the Bering Land Bridge and Cordilleran Ice Corridor were not passable until 9,000 years ago, an increasing amount of physical evidence indicates that man was surely in this hemisphere much earlier than that recent date.

For one thing, corn, a New World contribution to the pantries, is said to be, at 9,000 years, our oldest domesticated

seed crop. Some earliest agriculturalist had to be on this continent more than 9,000 years ago in order to domesticate the
seed. Conclusive proof that such ancient farmers did exist
was offered when a Humble Oil Company drill brought up
Mexican corn pollen that was more than 80,000 years old.

The anomalous Indian blood seration and dentition and the
geographic distribution of the American Indian demands an
impossible genetic time scale in which to transform Asiatic
immigrant to distinctive New World inhabitant. Even if we attempt to keep some kind of peace with the accepted theories
of New World habitation, we must grant more evolution in
40,000 years in North America than that which took place in
more than one million years in Europe, Africa, and Asia.

We might rest our case by providing as evidence the 50,000-
year-old skulls found in California, which are clearly those of
American Indians, but we are left with another mystery. A
140,000-year-old American Indian-type skull (via metric analysis) has been found at an Iranian excavation site.

What enormously complicates the above finding is the
uneasy fact that no precedent or prior skull types of the
American Indian have been found anywhere in the world. The
Amerindians just seem to appear without any evolutionary
transformational base.

The sites of the mysterious Mullions also offer no end of
trouble to conventional timetables and esteemed evolutionary
structures.

The Mullion culture suddenly appeared approximately
10,000 years ago along the Algerian coast with the largest
skeletal population in the entire prehistoric record. In addition, the Mullions also possessed the largest cranial capacity
of any population the world has ever known—approximately
2,000 cc versus our present 1,400 cc.

Whoever the Mullions were, they inhabited the site only
briefly, and their population consisted mostly of women and
children, who worked with tool types and domesticated animals never before seen.

As we shall see again and again, we have such a limited knowledge and practically no understanding of the worlds before our own. For example, who walked the streets of that remarkable "minicivilization" in Yugoslavia? Whoever strode the boulevards of those now silent cities were once citizens of a culture that flourished before ancient Egypt and China—5,000 years before the glory that was Greece.

What of the lost Amerindian civilization of Cahokia, complete with pyramids and a great wall? One site, near the present city of St. Louis, may have contained a metropolis of more than 250,000 North American Indians.

And who constructed the mysterious seven-mile walls of the Berkeley and Oakland, California, hills? Or the stone wall that runs for 20 miles near Petra in Jordan? And which pre-Mayan peoples engineered an elaborate waterworks in Yucatan to irrigate crops over 2,000 years ago?

And orthodox pooh-poohing to the contrary, the pyramids really do hold many mysteries. No one has ever really answered the anomalies of the Egyptian and Mexican pyramids; no one has truly dated them; and if we accept established theories for their construction, no one has explained how the earliest and smallest populations could erect the largest architecture.

Scientific knowledge has seemingly been prized by certain inhabitants of every culture, known and unknown.

Rock engravings, which may be as old as 60 million years, depict in step-by-step illustrations an entire heart-transplant operation and a Caesarean section. These 150,000 engraved stones also portray scenes indicating that a very special race of man actually lived at the same time as the prehistoric monster reptiles.

The ancient Egyptians used the equivalent of contraceptive jelly and had urine pregnancy tests.

The cement used in filling Mayan dental cavities still holds after 1,500 years.

Cosmas and Damian, two brothers, accomplished successful leg transplants 1,700 years ago. Had they somehow acquired knowledge of the techniques of gifted surgeons from the dim past?

The Caracol Tower at Chichén Itzá is a remarkable ancient Mesoamerican observatory that seems to have correlated its findings with similar sites in North America, including Mesa Verde, Wichita, and Chaco Canyon.

No fabric is supposed to have been found until Egypt produced cloth material 5,000 years ago. How, then, can we deal with the Russian site which provides spindle whorls and patterned fabric designs more than 80,000 years old?

Not only did the ancient Babylonians appear to use sulphur matches, but they had a technology sophisticated enough to employ complex electrochemical battery cells with copper wiring. There is also evidence of electric batteries and electrolysis in ancient Egypt, India, and Swahililand.

Recently interpreted ancient Aramaic texts suggest that the wandering Israelites were given a machine for producing the "manna from heaven" that kept them alive on their exhaustive desert trek.

There is evidence of a metal-working factory of over 200 furnaces which was found at what is now Medzamor in Russian Armenia.

Although a temperature of over 1,780 degrees is required to melt platinum, some pre-Incan peoples in Peru were making objects of the metal.

Even today the process of extracting aluminum from bauxite is a complicated procedure, but Chou Chu, famous general of the Tsin era (265-316 A.D.), was interred with aluminum belt fasteners on his burial costume.

Relics found embedded in solid rock or in ancient, undisturbed strata are called "erratics." The discoveries of these erratics suggest a vast antiquity for man's existence on this planet.

Carved bones, chalk, stones, together with what would appear to be greatly ornamented "coins," have been brought up from great depths during well-drilling operations.

A strange, imprinted slab was found in a coal mine. The artifact was decorated with diamond-shaped squares with the face of an old man in each "box."

In another coal-mine discovery, miners found smooth, polished concrete blocks which formed a solid wall. According to one miner's testimony, he chipped one block open only to find the standard mixture of sand and cement that makes up most typical building blocks of today.

A gold necklace was found embedded in a lump of coal.

A metal spike was discovered in a silver mine in Peru.

An iron implement was found in a Scottish coalbed, estimated to be millions of years older than man is believed to have existed.

A metal, bell-shaped vessel, inlaid with a silver floral design, was blasted out of solid rock near Dorchester, Massachusetts.

Two hypotheses may explain the presence of these perplexing artifacts, these "erratics": (1) that they were manufactured by an advanced civilization on Earth which, due either to natural or technological catastrophe, was destroyed before our world's own genesis; (2) that they are vestiges of a highly technological civilization of extraterrestrial origin, which visited this planet millions of years ago, leaving behind various artifacts.

This present volume shall emphasize the former theory. It will show that erratics do not appear to be the products of a "supercivilization" capable of interplanetary flight. These objects are generally of a workmanship which ranges in technical sophistication from what we would judge by our orthodox historical yardstick as comparable to that possessed by the societies of Greece or Rome or to that level achieved by our own twentieth-century technology.

Even if a highly advanced extraterrestrial race might have

visited this planet in prehistoric times, it seems unlikely that such common, everyday items as nails, necklaces, buckles, and vases would have been carried aboard a spacecraft and deposited in such widely separated areas; for erratics have been found in North and South America, Great Britian, the whole of Europe, Africa, Asia, and the Mid-East.

The question of just what exactly may have happened to these worlds before our own is one that invites a great deal of highly speculative thinking. In spite of the general unpopularity of catastrophism, there does seem to be a number of recently discovered "proofs" of ancient cataclysmic changes in the Earth's crust, which may account for the nearly total disappearance of these prehistoric worlds. Geological evidence indicates that these changes were both sudden and drastic and might have completely overwhelmed and destroyed the early inhabitants and their cultures.

Perhaps the most potentially mind-boggling evidence of an advanced prehistoric technology that might have blown its parent-culture away is to be found in those sites which ostensibly bear mute evidence of pre-Genesis nuclear reactions.

In an earlier work, I told of the "fused green glass" found deep in the strata of an archaeological dig. The statement that such material had been known previously only at nuclear testing sites (where the sand had melted to form the substance) proved to be an unsettling thought for many readers. But throughout the planet, in the same geologic strata, there unequivocally exists areas which scientists state are strongly suggestive of nuclear reactions. Could it be possible that these sites provide evidence of a prehistoric nuclear war?

"Fused green glass" has been found in such sites as Pierrelatte in Gabon, Africa; the Euphrates Valley; the Sahara Desert; the Gobi Desert; Iraq; the Mohave Desert; Scotland; the Old and Middle Kingdoms of Egypt; and south-central Turkey. At the same time, scientists have found a number of uranium deposits that appear to have been mined or depleted in antiquity.

But before we confront the grim specter of nuclear annihilation in prehistoric times and begin to worry about history repeating itself in some kind of cyclic purgation, let us attempt to meet head on the mystery of mankind's genesis. It would seem only proper, after all, to resolve its beginnings before dealing with its ending.

CHAPTER TWO

The Genesis Mystery

"The study of prehistory today is in a state of crisis," Colin Renfrew writes in the introduction to his *Before Civilization.*"Archaeologists all over the world have realized that much of prehistory, as written in the existing textbooks, is inadequate: some of it quite simply wrong. A few errors, of course, were to be expected, since the discovery of new material through archaeological excavation inevitably leads to new conclusions. But what has come as a considerable shock, a development hardly foreseeable just a few years ago, is that prehistory as we have learnt it is based upon several assumptions which can no longer be accepted as valid."

Renfrew considers these revolutionary changes so threatening to the fundamental view of the past that scientists must now begin to shift to a new paradigm, an entirely new framework of thought.

For instance, nearly all students of prehistory have been taught that the pyramids of Egypt are the oldest stone-built monuments in the world and that the first elaborate places of worship constructed by man were to be found in the land of Mesopotamia. There in the Near East, metallurgy was born, together with architecture and other cultured skills. Civilization flowed with its fertile, domesticating strengths over Europe and Great Britian.

Now, bemoans Renfrew, we have been shocked to learn that all of these assumptions are incorrect: "The megalithic chamber tombs of western Europe are now dated earlier than the Pyramids. . . . The impressive temples of Malta are now set before any of their Near Eastern counterparts in stone. Copper metallurgy appears to have been underway in the Balkans at an early date—earlier than Greece—so that it may have developed quite independently in Europe. And Stonehenge was, it seems, completed, and the rich early bronze age of Britian well under way, before the Mycenaean civilization of Greece even began. In fact Stonehenge, that remarkable and enigmatic structure, can now be claimed as the world's oldest astronomical observatory. The traditional view of prehistory is now contradicted at every point."

Perhaps nowhere do the traditional views of prehistory seem more contradicted and confused than in the area of determining modern man's genetic antecedents. The Genesis Mystery is a detective melodrama in which an incredible array of fantastic characters seem always to be popping up with false clues and in which even one's most trusted and apparently reliable allies later reveal themselves as woefully undependable. Packing crates full of new evidence indicate that modern man is much older than the academics previously believed and that sophisticated civilizations developed much earlier than the orthodox timetables had previously permitted. And with the discovery of traces of man dating back to 70,000 B.C. in southern California, it may soon be only the intellectual laggards who will persist in terming the Americas the "New World."

The most deeply entrenched theory of man's origins, which

has him being rocked to adolescence in the Middle East's "cradle of civilization," is under severe attack by recent discoveries of early metal alloys and pottery in Thailand. The bronze artifacts, which date back to 3600 B.C., "challenge all assumptions that have long been held about the development of our modern cultures," according to one expert. The pottery fragments are believed to be 600 years older than similar workmanship representative of the Tigris-Euphrates valley, thereby demonstrating that Asia may have given pottery-making and metal-working skills to the Middle East, not vice versa, as has so long been maintained.

Western archaeologists have begun to prod the strata of eastern Africa for evidence that the earliest primate originated there between two and five million years ago. In August of 1976, Tanzanian officials announced that the skull of a "missing link" creature had been found at Lake Ndutu. According to one official, "The cranium is remarkable in that it seems to form an evolutionary link between [Peking man] and *Homo sapiens* [humans], having features in common with both."

Although the Ndutu man was found near materials carbon-dated at about 500,000 years old, Chinese scientists had declared in July, 1976, that they had found teeth and stone implements which proved that the "Yuanmo man" lived in what is now Yunnan province more than 1.7 million years ago. The New China News agency said, "This dating pushes the age of the earliest apeman discovered in China back more than one million years." *Red Flag* magazine, the theoretical journal of Chinese Communism, stated: "It is now established that a fairly long period of time separated the time the ape began to make tools from the 'Peking era' in the evolution toward man."

It seems as though previously undiscovered "lost civilizations" are being unearthed with incredible frequency. On November 28, 1976, ruins near La Paz, Bolivia were offered to public scrutiny for the first time. Carlos Once Sangines, national director of archaeology in Bolivia, said that the Mollo culture used the trapezoidal shape in architectural works that predate the Incan empire. The Mollo peoples amassed their vast kingdom in the Andes between the thirteenth and fifteenth centuries before the Incas. Thus, while the trapezoid has long been considered an Incan innovation, it has now been revealed that an earlier people had already discovered the secret of the magnificent, massively built structures that have puzzled archaeologists for centuries.

Italian archaeologist Paulo Matthaie chose Syria as his hunting place for a prehistoric Shangri-La. In the northern section of a country that has long been dismissed as nothing more than a land of nomads devoid of culture or commerce, Matthaie and his crew found nearly 15,000 tablets in the royal palace of the previously unknown Elba kingdom. The tablets comprise a history of events from 2500 B.C. to 2400 B.C. and were written in a script very similar to biblical Hebrew, which would not develop for another several hundred years.

Matthaie believes the tablets provide "evidence of the existence of a new world that rivaled the ancient kingdoms of Egypt and Mesopotamia" and will offer an "important new chapter in the history of the world."

The Corozal Project, joint venture of the British Museum and Cambridge University, has been excavating a number of Maya ceremonial centers since 1973. One of their expeditions investigated a Mayan monument which bears what may be the oldest recorded date yet found in the New World, "no later than the first century B.C. and possibly as much as one to two hundred years earlier."

A sample of burned wood from a site in Cuello, Belize, was radiocarbon dated at 2600 B.C. The researchers state in the April 15 issue of *Nature* that this determination places "the origins of Maya settlement and civilization in the Yucatan Peninsula back in the third millenium B.C., some 1,700 years earlier than the first occupation known until now."

Homo erectus, the familiar Peking man and Java man, was originally dated at about 500,000 years old and was deemed to be our earliest ancestor. Erectus' age was pushed back to something over a million years by discoveries made at Olduvai Gorge in Tanzania in 1960. Then in August, 1972, Richard Leakey's young associate Bernard Ngeneo found a shattered cranium in a steep gully in the gray-brown wastelands east of Kenya's Lake Rudolf that may shatter any present form of rigid thinking in regard to man's pedigree.

"Either we toss out this skull or we toss out our theories of early man," Leaky declared of the 2.8-million-year-old fossil that he tentatively identified as belonging to the genus of modern man.

"It simply fits no previous models of human beginnings," Leakey continued in an article in the June, 1973 issue of *National Geographic.* The skull's surprisingly large braincase, according to Leakey, "leaves in ruins the notion that all early fossils can be arranged in an orderly sequence of evolutionary change. It appears that there were several different kinds of early man, some of whom developed larger brains earlier than had been supposed."

The Leakey researchers named our unknown cousin "1470 man," after the registration number assigned to the specimen by the National Museum of Kenya.

"It became apparent that the skull lacked the protruding eyebrow ridges—beetle brow—of *Homo erectus*," Leakey commented. "And the braincase, though nearly three times as old as *Homo erectus*, was nearly as large. . . . In the laboratory Dr. Alan Walker . . . confirmed our estimate of 800 cubic centimeters. By comparison, skull specimens of the much later *Homo erectus* range between only 750 and 1,100 cc. (The average modern human cranium holds about 1,400 cc.)"

Richard Leakey's discoveries have convinced him that there may have been a number of models of early man—"geographical or regional variants of the same species." Leakey feels confident that anthropologists will one day be able to "follow man's fossil trail at East Rudolf back as far as four million years. There, perhaps, we will find evidence of a common ancestor for *Australopithecus*—near-man—and the genus Homo, true man."

On October 17-18, 1974, a French-American expedition led by D. Carl Johanson of Case Western Reserve University in Cleveland helped a four-million-year-old-man out of his volcanic grave. The remarkable distinterrment not only threatens to destroy all current theories on the origin of our species, but it re-establishes the Middle East as the birthplace of man.

Johanson recalled that the researchers had been giddy with excitement over finding a jawbone of an extinct hyena, when Alemeyu Asfew of the Ethiopian Antiquities Administration came running over a small hill. "He was so excited he practically couldn't talk," Johanson said. "He had found a palate and teeth of a human more than three million years old."

The crew went on to uncover a complete upper jawbone, a half upper jaw and a half lower jaw, all with teeth intact. Preliminary dating disclosed that the fragments may be four million years old. In the matter of two days, Johanson said, "We have extended our knowledge of the genus Homo by nearly 1.5 million years."

Although the fossils were found in the Afar region of northeastern Ethiopia, the new find indicates the Middle East, not Africa, as the birthplace of man, according to Johanson. The four-million-year-old fragments were discovered on the surface of a volcanic deposit on the Hadar, a tributary of the Awash River, only 100 miles from the Red Sea, where Africa and Arabia were once connected by a land bridge.

"The small size of the teeth in these jawbones leads us to hypothesize that the genus Homo was eating meat and probably using tools, perhaps bones, to kill animals three to four million years ago," Johanson stated. "It also means that there was probably some kind of social cooperation and some sort of communication system."

Science Digest (February, 1975) reported: "The bones lay in a stratigraphic level 150 feet beneath a volcanic layer which has been dated between 3.0 and 3.25 million years old, thus lending credibility to Johanson's claim they are close to four million years old."

While certain orthodox archaeologists and anthropologists find themselves embroiled in heated controversies when they attempt to push the date of man's genesis beyond a million years, irritating erratics continue to be unearthed which suggest a vast antiquity for the genus Homo. And they have been turning up and being overturned for generations now. Witness this letter which appeared in the March 27, 1873 issue of *Nature*:

> . . . Mr. Frank Calvert has recently discovered, near the Dardanelles, what he regards as conclusive evidence of the existence of man during the Miocene period. Mr. Calvert had previously sent me some drawings of bones and shells from the strata in question, which Mr. Buck and Mr. Gwyn Jeffreys were good enough to examine for me. He has now met with a fragment of bone, probably belonging either to the Dinotherium or a Mastodon, on the convex side of which is engraved a representation of a horned quadruped, "with arched neck, lozenge-shaped chest, long body, straight forelegs, and broad feet." There are also, he says, traces of seven or eight other figures, which, however, are nearly obliterated. He informs me that in the same stratum he has also found a flint flake, and several bones broken as if for the extraction of marrow.
>
> This discovery would not only prove the existence of man in Miocene times, but of man who had already made some progress, at least, in art. Mr. Calvert assures me that he feels no doubt whatever as to the geological age of the stratum from which these specimens were obtained. . . . JOHN LUBBOCK

The Miocene is a substratum of the Tertiary period, which comprises the geologic time of roughly 100 million years ago. In his *Fossil Man*, Frank Cousins discusses the human remains found in Castenedolo and Olmo, Italy, which ostensibly bear additional evidence for man in Tertiary times.

In 1860, Professor Ragazzoni, an expert geologist and a teacher in the Technical Institute at Brescia, found the fragmentary vault of a human skull in deposit of coralline stratum of the Pleistocene glaciations (circa ten million years old). He searched farther and found a few other cranial fragments. When he later showed his finds to his colleagues at the Institute, they were received with the utmost incredulity.

Twenty years later, a friend of Ragazzoni's, excavating in the same pit in which the cranial fragments were discovered, found the scattered skeletal remains of two children. The skeletons were left *in situ* until they could be seen and examined by Professor Ragazzoni. Later, the skeleton of a woman in the contracted posture was found within the same stratum.

In 1883, an anthropologist of rising reputation, Professor Sergi, visited Ragazzoni at Brescia and examined the human remains which had been found in the Pliocene strata at Castenedolo. The fragments were still covered by the original matrix in which they had been embedded, and Professor Sergi pronounced the remains to be those of two children, a man, and a woman of the modern type.

The anthropologist accompanied Ragazzoni to the pit from which so much had been surrendered to the curious scrutiny of contemporary man; and he, himself, made a fresh section of the strata. He was convinced that Ragazzoni had in no way misrepresented his startling discoveries, namely, that human remains had been found in undisturbed beds of a Pliocene age, and that these fragments represented a species of man of the modern type.

In 1863, while making the railroad southward from Arezzo, in the upper waters of the Arno River, a trench over 50 feet deep had to be dug. It was during this excavation that the Olmo skull was unearthed.

I. Cocchi, curator of the Museum of Geology in Florence, said that the skull lay at a depth of nearly 50 feet beneath the surface in a deposit that had been formed in the floor of an ancient lake. The blue clay in which the skull was found was determined by Signor Cocchi to be of older Pleistocene deposits. The remains of an elephant and an early form of Pleistocene horse were discovered at the same level as the human skull.

Then there are those truly annoying reports of human remains being located in coal beds. If there were man creatures existing in the Carboniferous period, the geologic time strata in which our massive coal beds were formed, we must be talking about an ancestor of modern man existing over 600 million years ago. Before we more closely examine the question in the next chapter, here is a "for instance," from *Geology of Coal* by Otto Stutzer:

> Animal remains in coal beds are extremely rare. The animals which once inhabited the great coal swamps were terrestrial forms, the bodies of which decomposed after death just as rapidly as do the bodies of animals living in existing primitive forests and moors. In the coal collection of the Mining Academy in Freiberg there is a puzzling human skull composed of brown-coal and manganiferous and phosphatic limonite, but its source is not known. This skull was described by Karsten and Dechen in 1842.

It soon becomes clear to even the most rational of scholars that our Genesis Mystery rapidly deteriorates into a presently hopeless muddle of conflicting data and controversial claims. Man's family tree obviously has more limbs than any professional anthropologist cares to explore. After all, tenure could come quickly crashing down around the ears of the venturesome academic if some more conservative colleague chose to saw away at the limb on which he was building his avant-garde reputation.

At the present time, we might assemble a data sheet regarding man's origins that would look something like the following:

A consensus among those scientists concerned with the Genesis Mystery holds that modern man, *Homo sapiens*, became the dominant species about 40,000 years ago and has been on Earth for about 80,000 years.

Cro-Magnon, a tall, handsome, prehistoric race of Europe, is believed to be of the same species as modern man and may have been absorbed by *Homo sapiens*.

Neanderthal, classified as *Homo sapiens,* was extant from 150,000 down to 50,000 years ago.

Other sapiens fossils have been found that suggest a greater antiquity for the species. Those found at Swanscombe in England and Stienhiem in Germany are believed to be 250,000 years old. A fossil find in Hungary is, according to some researchers, 500,000 years old.

Fossils belonging to the genus homo, but not the modern species, sapiens, are called *Homo erectus*. Within this classification are the fossil find at Heidelberg, Germany, dated at 350,000 years; the discoveries in China (*Sinanthropus*), 400,000 years; those in Java, called *Pithecanthropus*, 400,000 to 700,000 years.

Those who do not qualify for the genus Homo, but are manlike and therefore within the family of Homoinidae possibly in the evolutionary line leading to modern man, are the *Australopithecines*, including Dr. Louis Leakey's *Zinjanthropus*, placed at 1,750,000 years. Dr. Leakey also championed *Homo habilis* as a Hominidae of the same period.

Richard Leakey's "1470 man," which he tentatively

identified as belonging to the genus Homo, dated at 2.8 million years old, would probably not be widely accepted among the majority of professional-origins scientists.

Nor would Dr. Johanson's four-million-year-old Ethiopian immigrant from the Middle East.

Richard Leakey foresees that one day anthropologists will be able to find a four-million-year-old common ancestor for near-man and the genus Homo, true man. His would seem to be a minority vision at this time.

The February 13, 1967 issue of *Newsweek* magazine declared: "The evidence for man's evolution could hardly be more tenuous: a collection of a few hundred fossilized skulls, teeth, jawbones and other fragments. Physical anthropologists, however, have been ingenious at reading this record—perhaps too ingenious, for there are almost as many versions of man's early history as there are anthropologists to propose them."

Newsweek summarizes the few facts on which nearly all the scientists have agreed: "The generally accepted figure for creatures with upright posture and manlike teeth is 1.7 million years. . . . The first appearance of the Hommidae, a family distinct from the apes, whose sole surviving member is modern man . . . apparently emerged about 1.4 million years ago. . . .

In the next chapter we shall be pointing to some fantastic fossils which would seem to indicate that modern man not only began, but flourished, in an incredible world that existed long before our own.

CHAPTER THREE

Footprints in the Stones of Time

It is as if these impossible fossils serve as a kind of goad for mankind's collective unconscious. It is as though these incredible footprints in the stones of time remain as an irritant that the blinking eye of man's intellectual curiosity can never quite remove.

There seems to be an inordinate number of these fantastic fossils in the United States, and our scientists have never totally ignored them (nor, of course, have the majority of our researchers and academics accepted them).

The enchanted mountain, about two miles south of Brass town [Tennessee] is famed for the curiosities on its rocks. There are on several rocks a number of impressions resembling the tracks of turkeys, bears, horses, and human beings, as visible and perfect as if they were made on snow or sand. *American Encyclopedia*, Dobson, Philadelphia, 1778-1803; Supplement, vol. 3, p. 344.

[From an unnamed, but "eminent," English geologist, dated

September 9, 1837] . . . In the 5th volume of your Journal (1822), there are remarks on the prints of human feet observed in the secondary limestone of the valley of the Mississippi, by Mr. Schoolcraft and Mr. Benton, with a plate representing the impressions of two feet. Ever since my researches on the rippled sandstones (published in Jameson's *Edinburgh Journal*), I felt persuaded the prints alluded to were the genuine impressions of human feet, made in the limestone when wet. I cannot now go on with the arguments that may be urged in proof of my opinion; but rely upon it, those prints are certain evidence that man existed at the epoch of the deposition of that limestone. . . . I am prepared to find man and the contemporary animals much lower down in the series than is generally supposed. My friend Sir Woodbine Parish (the discoverer of the Megatherium), tells me that similar impressions have been seen in South America; and there was a dispute among the catholics whether they were the feet of the apostles! *American Journal of Science,*Vol. 33, 1838.

It is reported that H. E. Huford of Kemper Lane, Walnut Hills, Cincinnati, exhibited before a recent meeting of the Ohio State Academy of Science, a large stone taken from the hillsides four miles north of Parkersburg, on the West Virginia side of Ohio river, about twenty years ago, in which there was the imprint of a perfect human foot, 14½ inches in length. The matter will be investigated by the Society. *American Anthropologist,*Vol. IX, 1896.

Scientifically inclined persons in Southern California are pondering deeply over the discovery, in Elysian Park, of a distinct imprint in solid stone of a shoe worn by a human being, says the Los Angeles *Herald*. This discovery is certain to excite no little comment, for there appears to be no authentic record of a fossilized footprint of a human being of ordinary size having been found heretofore.

The peculiar feature of this find is that the owner of the foot wore a shoe of antique Mexican fashion, with high, narrow heel and broad, flat sole. The imprint is perfectly clear and looks as though the owner had unwittingly put his right foot into soft mud but a day or two ago and left his mark. . . . The fossil imprint was discovered by laborers who were making a deep cut for the new wagon road northeast of the park. It was

cut out of solid rock, four feet or thereabouts below the surface soil and from a point on the hillside at least 70 feet above
the bottom of the little canyon at the hill's base. The stone is a
fine-grained shale, impregnated with lime. . . .

. . . Several other excellently preserved and clearly deli-
neated organic remains, such as ferns, leaves, and twigs, have
been found in the same deposit, and, stranger still, but a few
days ago, the complete outline of a fish was taken from the
same situation of stone, not many feet distant from this last
find. The fish's remains were twenty feet below the surface.
The Morning Star, Savannah, Georgia, April 13, 1897.

As we have noted in the preceding chapter, man is supposed to have evolved only in the late Tertiary period and is
therefore only about one million years old. But fossilized
footprints have been found in rocks from the Carboniferous
period to the Cambrian period, thus offering mute, but dramatic, testimony that some bipedal creature was walking
about from 250 to 500 million years ago.

The fossil tracks of both bare and shod feet of a decidedly
human impression have been found in sites ranging from Virginia and Pennsylvania, through Kentucky, Illinois, Missouri,
Utah, Oklahoma, and Texas. The prints give every evidence
of having been made by human feet at a time when the rocks
were soft mud or pliant sand.

Although the discovery of these footprints in the stones of
time are hardly rare or recent occurrences, geologists by and
large refuse to accept the fossil evidence at face value because to do so would be to acknowledge that modern man
lived in the earliest years of hypothetical evolutionary history. Albert C. Ingalls, writing of such footprints as "the carboniferous mystery" in *Scientific American*, January, 1940,
states:

If man, or even his ape ancestor, or even that ape ancestor's
early mammalian ancestor, existed as far back as in the Carboniferous Period in any shape, then the whole science of geol-

31

ogy is so completely wrong that all the geologists will resign their jobs and take up truck driving. Hence for the present at least, science rejects the attractive explanation that man made these mysterious prints in the mud of the Carboniferous Period with his feet.

If the scientists can dismiss the footprints as tracks made by some as yet undiscovered Carboniferous amphibian that walked upright and left no dragging tail or sliding belly marks, it would seem that human teeth, bones, and artifacts in Carboniferous deposits must present much larger lumps to sweep under the academic carpet.

In 1912, two employees of the Municipal Electric Plant, Thomas, Oklahoma, used a sledge to break apart a chunk of coal too large for the furnace. An iron pot toppled from the center of the lump, leaving an impression in the coal. The coal had been mined near Wilburton, Oklahoma, an area of the southwest that seems to yield numerous erratics and anomalous footprints. The two men freely signed an affidavit attesting to their incredible discovery. The artifact has been photographed and thousands of curious men and women have examined the pot from an unknown time and place.

On November 7, 1926, fossil hunters in the coal beds of the Bear Creek Field near Billings, Montana, found a human tooth, the enamel of which had long since turned to carbon and the lime of the roots to iron. According to *The New York Times* for November 8, 1926:

> The tooth, declared by dentists of this city [Billings] to be the second lower molar of a human being, was found by Dr. J. C. Siegfriedt of Bear Creek, who has been collecting fossils for the University of Iowa and for other institutions.
>
> The coal deposit is in the fortunian formation, lowest of those laid down in the Eocene Period. Many fossils, including ganoids, a kind of fish scale, and sharks' teeth have been found in the deposit by Dr. Siegfriedt, who asserts that it furnishes prolific material for fossil and dinosaur research.

32

In 1958, Professor Johannes Hurzeler of the Museum of Natural History in Basel, Switzerland, found the jawbone of a child flattened like a piece of sheet iron in a lump of coal dated from the Miocene age, roughly ten million years ago.

In private correspondence, W. W. McCormick of Abilene, Texas, relates his grandfather's account of a building that was buried in a coal mine shaft:

> In the year 1928, I, Atlas Almon Mathis, was working in coal mine No. 5, located two miles north of Heavener, Oklahoma. This was a shaft mine, and they told us it was two miles deep. The mine was so deep that they let us down into it on an elevator. . . . They pumped air down to us, it was so deep. . . .
>
> . . . One night I shot four shots [referring to blasting coal loose] in room 24 of this mine, and the next morning there were several concrete blocks laying in the room. These blocks were 12-inch cubes and were so smooth and polished on the outside that all six sides could serve as mirrors. Yet they were full of gravel, because I chipped one of them open with my pick, and it was plain concrete inside.
>
> As I started to timber the room up, it caved in; and I barely escaped. When I came back after the cave-in, a solid wall of these polished blocks was left exposed. About 100 to 150 yards farther down our air core, another miner struck this same wall, or one very similar. Immediately they [the mining company officers] pulled us out of this wing of the mine and forebade us to tell anything we had seen.
>
> This mine was closed in the fall of 1928, and the crew went to Kentucky . . .
>
> . . . Before I started working on this crew, they had a similar experience in mine 24 at Wilburton, Oklahoma in about the year 1926.
>
> They said they dug up two odd things: One was a solid block of silver in the shape of a barrel, and the other was a bone that was about the size of an elephant's. I don't know if they meant only in diameter or in diameter and in length, but they did say it had knuckles on each end. The silver block had the prints of the staves on it, and the saw that first struck it cut off a chip on the edge at one end. The miners saw the silver dust the saw was pulling out and went to dig out the block.

What was done with these things, I do not know. In the case
of the blocks in my room in No. 5, I don't think any were kept.

A few years ago Dr. Henry Morris, an enthusiastic support-
er of the Creationist movement, reported that he had personally interviewed a coal miner in West Virginia who had excavated a perfectly formed human leg that had changed into
coal. He also encountered the claim that miners in the general
area of West Virginia had unearthed a well-constructed concrete building.

In May of 1971, while searching for little spheres of blue
azurite on a "rockhound" tour of bulldozed land that belonged to a mining company, Lin Ottinger, tour guide and
amateur geologist-archaeologist, found trace of human remains in a geological stratum that is approximately 100 million
years old.

The find was precipitated when a woman collector handed
Ottinger a specimen for identification. He immediately recognized the object as a human tooth.

Ottinger shouted, assembling the group of scattered rockhounds he had brought to the Big Indian Copper Mine in Lisbon Valley, about 35 miles south of Moab, Utah. After he had
told the men and women what to look for, Ottinger joined
them in a careful, coordinated search of the bulldozed area.

Might it be possible that there could also be bits of bone as
well as additional teeth lying half-buried after their rude disin-
terment by the mining company's bulldozers? Ottinger in-
structed his rockhounds to look carefully for the telltale
brownish stains that decaying organic matter leaves in sand.

In a matter of a few minutes, Ottinger's spontaneously assembled crew had located several more teeth and a number of
bone fragments, one one of which was quite obviously from a
human jawbone.

Then someone gave a triumphant cry of discovery. A trace
of brown discolored the white, semi-rock sand.

Ottinger knelt above the dark stain and carefully began probing the decomposing sandstone with his knife blade. With almost surgical skill, Ottinger soon uncovered a smoothly rounded bone that had acquired a greenish hue from the copper in the sandstone.

At this point, the amateur archaeologist demonstrated his professional attitude toward what could be a major find. He stopped digging. He knew that human bones left in place, *in situ*, as he found them in ancient rock strata could present the ax that could topple the current consensus of how long man's family tree had been growing.

If one finds human bones *in situ* in a rock formation, then the skeletal fragments have to be as old as the rock that surrounds them. Of course the very idea of human bones in rock is repellent to the ears of any orthodox scientist, for his dogma declares that the genus Homo and any of his ancestors is far younger than the very newest rock formations.

Lin Ottinger carefully covered the exposed fragments with moistened paper, then sprinkled the paper with loose sand. He wanted to be certain that the bones would be protected from exposure to the desert air.

Ottinger was well aware of the fact that only an accredited scientist could establish whether or not the bones were truly human and if they were "in place" within the rock, so he notified Dr. W. Lee Stokes, with whom he had worked on previous paleontological finds.

According to the Moab, Utah, *Times Independent*, June 3, 1971: "The implications of Ottinger's find were instantly recognized by Dr. Stokes. If the human remains were truly "in place" in the Dakota formation, that is, not washed or fallen in from higher and younger strata, then the remains would have to be the same age as the stratum in which they were found. This would be in the vicinity of 100 million years. . . ."

Dr. Stokes referred the investigation to a colleague, Dr. J.

P. Marwitt, professor of anthropology at the University of Utah. A natural history television photography team, a local news reporter, and a number of interested people accompanied Ottinger and Dr. Marwitt to the desert valley site.

Dr. Marwitt immediately set to work uncovering the bones.

"Parts of at least two separate skeletons were exposed in this preliminary survey," reported the *Times Independent*. "While Marwitt and Ottinger were working on the prime site, several volunteers were screening loose sand and dirt in the vicinity for teeth and bone shards. Quite a number were found."

As the skeletons were uncovered, it soon became apparent that they were "in place" and had not been washed in or fallen down from higher strata.

"The portions of skeletons that were exposed were still articulated, that is, were still joined naturally, indicating that the bodies were still intact when buried or covered in the Dakota formation," the *Times Independent* stated.

An interesting side feature of the find was the fact that the bones had been stained a bright green by the copper salts that occur in the vicinity.

"In addition," the *Times Independent* commented, "the dark organic stains found around the bones indicate that the bones had been complete bodies when deposited in the ancient stratum."

Dr. Marwitt pointed out a number of curious aspects of the remarkable find. One of the bodies appeared to be in the position very often used by ancient Indian tribes in their formal burial observances, but the upper body of the other skeleton had been carried away. The bulldozer that had removed the rock and other materials from the site was blamed as the most likely body snatcher.

Mine metallurgist Keith Barrett remembered that the rock and soil that had been above the remains before the bulldozer work had begun had been "continuous . . . with no caves or major faults or crevices visible. Thus, before the mine exploration work, the human remains had been completely covered by about fifteen feet of material, including five or six feet of solid rock. This provided strong, but not conclusive, evidence that the remains are as old as the strata in which they were found.''

And, once again, we are speaking of an age of at least 100 million years. Due to a certain local faulting and shifting, the site could be either in the lower Dakota or the still older upper Morrison formation.

Scientists, though, found a serious contradiction inherent in the find. Even though the skeletons were found in rock stratum over 100 million years old, they appeared to be the remains of *Homo sapiens*, modern man, not some ancient, apelike predecessor.

''. . . Even though the rock and soil layers originally above the bones were continuous and unbroken as claimed by mine officials, there is still the possibility, in fact a high probability, that the original owners of the bones had simply been using a narrow cave in the Dakota formation, when it collapsed and buried them, then later filled in solid with the sandy soil that surrounded the bones when they were found,'' stated the *Times Independent* in summation of the scientist's position.

Laboratory age-dating seemed to be the only method of resolving the mystery presented by the human bones in rock, ostensibly over 100 million years old. Dr. Marwitt removed the skeletal fragments and transported them to his university's laboratories.

And it is here, according to F. A. Barnes's article in the February, 1975 issue of *Desert* magazine, that the matter rests to this day:

"Somehow, the university scientists never got around to age-dating the mystery bones. Dr. Marwitt seemed to lose interest . . . then transferred to an eastern university. No one else took over the investigation. Lin Ottinger, growing tired of waiting after more than a year, reclaimed his box of bones.

". . . It is highly probable that the bones, are, indeed, this old. Yet, who knows? Without that vital age-dating, no one can say positively that they are not . . . .

". . . Part of the mystery, of course, is why the University of Utah scientists chose not to age-date the . . . bones and clear up at least the question of their actual age.

"And so the mystery remains, perhaps never to be solved."

And when will we be able to stamp "solved" across the dossier that contains the data concerning the discovery made on January 25, 1927 in Nevada of a shoe sole that was fossilized in Triassic limestone, thereby placing man back in the time of the giant reptiles? In the late 1920s, the Oakland Museum in California published a small bulletin under the title "The Doheny Scientific Expedition to the Hava Supai Canyon, Northern Arizona."

A Mr. Knapp, the discoverer of the erratic, writes that the fossil lay among some loose rocks. He picked it up and, upon later examination, "came to the conclusion that it is a layer from the heel of a shoe which had been pulled from the balance of the heel by suction; the rock being in a plastic state at that time. I found it in limestone of the Triassic Period, a belt of which runs through that section of the hills."

The relic was taken to New York where it was analyzed by a competent geologist of the Rockefeller Foundation, who verified Mr. Knapp's assessment and pronounced the fossil as unquestionably formed in Triassic limestone. Excerpting from the Oakland Museum's bulletin:

> Micro-photographs were made which showed very clearly that it bore a minute resemblance to a well-made piece of leather, stitched by hand, and at one time worn by a human foot. The photographs showed the stitches very plainly; at one place it was double-stitched and the twist of the thread could be clearly seen. The thread is smaller than any used by shoemakers of today. Minute crystals of sulphide of mercury are to be

noticed throughout the spaces of this fossil shoe sole, these minerals having been deposited in the long ago by waters which carried them in solution.

Samuel Hubbard, honorary curator of archaeology of the Oakland Museum, is quoted as saying: "There are whole races of primitive men on earth today, utterly incapable of etching that picture or sewing that moccasin. What becomes of the Darwinian Theory in the face of this evidence that there were intelligent men on earth millions of years before apes are supposed to have evolved?"

The White Sands National Monument near Alamogordo, New Mexico, contains some 176,000 acres of white alabaster. Geologists theorize that this gypsum was precipitated as arid winds dried up an inland sea. Somewhere in the great expanse of gypsum are what appear to be the sandal prints of some prehistoric human giant, who could only have made such impressions when the muddy sediment of the primeval ocean was beginning to harden.

In the *Story of the Great White Sands*, a booklet distributed at the National Monument, an account is related concerning the discovery of the massive human tracks:

> In the fall of 1932, Ellis Wright, a government trapper, reported that he had found human tracks of unbelievable size imprinted in the gypsum rock on the west side of White Sands. At his suggestion a party was made up to investigate. Mr. Wright served as guide. . . .
>
> As Mr. Wright reported, there were thirteen human tracks crossing a narrow swag, pretty well out between the mountains and the sands. Each track was approximately 22 inches long and from eight-to-ten inches wide. It was the consensus that the tracks were made by a human being, for the print was perfect, and even the instep plainly marked. However, there was no one in the group who cared to venture a guess as to when the tracks were made, or how they came to be of their tremendous size. It is one of the great unsolved mysteries of the Great White Sands.

Like so many really good mysteries, Ellis Wright's discovery of giant manlike tracks appears to have become more complicated over the years. In her "Happenings—Past and Present" column in the *Silver City Enterprise*, April 1, 1971, Mary Wright told of certain people who had contacted a "Ranger Bozart" for a tour of the area where Wright (apparently no relation to Mary) had discovered the alleged tracks of a prehistoric man. According to Ms. Wright, additional imprints in the gypsum had been found:

> . . . The party found more tracks which were going in the same direction as the first ones found, and each appeared to be using a cane.
>
> Since these tracks, which were in hardened caliche, were twice the size of the ordinary man's track, who were these early day travelers and what could they have been seeking in the San Andreas mountains?
>
> Some of the answers that may come out of the proposed study of this new find may be interesting. What were these early day travelers like? They must have been large; their tracks were twice that of the average man; and from the first pictures taken, they were wearing some type of sandal or moccasin. They crossed these lakes when the caliche was soft as their tracks show. They were headed west toward the mountains and can be tracked quite a distance before sand has covered the trail.

The sands of time have surely covered and obscured another trail that has lead many researchers into a steaming jungle of controversy over whether or not man lived at the time of the great reptiles more than 70 million years ago. To pursue such inquiry is to invite charges of being either a fantasist or a fanatic. But as we shall see in the next chapter, there are a number of rather learned individuals who feel that the evidence is far more than circumstantial that there were giant men on the Earth in those days who battled for survival against the giant reptiles.

CHAPTER FOUR

Giant Men and Giant Reptiles:

Did the Twain Meet?

From the mouth of the Illinois River at Grafton to Alton (Illinois), a distance of twenty miles, the Mississippi River runs from west to east, and its north bank (the Illinois side) is a high bluff. When the first white men explored the area, they found that some unknown muralist from some forgotten tribal culture had engraved and painted hideous depictions of two gigantic, winged monsters. The petroglyphs were each about thirty feet in length and twelve feet in height.

Father Marquette, the celebrated Jesuit priest-explorer, wrote in his journals of discoveries of the Mississippi, published in Paris in 1681: "As we were descending the river we saw high rocks with hideous monsters painted on them and upon which the bravest Indian dare not look. They [have] head and horns like a goat; their eyes are red; [they have a] beard like a tiger's and a face like a man's. Their tails are so long that they pass over their bodies and between their legs under their bodies, ending like a fish's tail. They are painted red, green, and black, and so well drawn that I could not believe they were drawn by the Indians, and for what purpose they were drawn seems to me a mystery.''

In a small volume published in 1698, Father Hennepin, another early explorer of the wilds of the west, wrote: ''. . . The Illinois [Indians] told us likewise that the rock on which these dreadful Monsters stood was so steep that no man could climb up to it, but had we not been afraid of the Savages more than of the Monsters we had certainly got up to them. . . . While I was in Quebec I understood M. Jolliet had been upon the [Mississippi] and obliged to return without going down the River because of the Monsters I have spoken of who had frightened him. . . .''

The two enormously large petroglyphs were clearly visible on the north bank of the Mississippi, immediately where the Illinois State Prison was later built at Alton. Traces of their outlines remained until the limestone on which they had been engraved was quarried by the convicts in about 1856.

In his forty-eight-page booklet, *The Piasa or The Devil Among the Indians* (Morris, Ill., 1887) P. A. Armstrong described the creatures as having ''. . . the wings of a bat, but of the shape of an eagle's. . . . They also had four legs, each supplied with eagle-shaped talons. The combination and blending together of the master species of the earth, sea, and air . . . so as to present the leading and most terrific characteristics of the various species thus graphically arranged, is an absolute wonder and seems to show a vastly superior knowledge of animal, fowl, reptile, and fish nature than has been accorded to the Indian.''

It is interesting to note that the petroglyphs were painted in only the colors red, black, and green: red, representative of war; black, symbolic of death and despair; green, expressive of hope and triumph over death in the land of dreams, beyond the evening star, where lie the grounds of peace and plenty.

Whatever the petroglyphs truly represented, all the Amerindian nations of what then constituted the Northwest Territory had a terrible tradition associated with the creatures they called *The Piasa* (or Piusa).

Sometime in the 1840s, Professor John Russell of Jersey County, Illinois, explored the caves which the Piasa were said to have inhabited and reported "innumerable human bones littering the stone floors." Although Professor Russell suggested the skeletal fragments offered mute testimony to the Amerindians' account of a flying monster with a craving for human flesh, P. A. Armstrong cautioned his readers that the cave may have been utilized as a burial place by the Mound Builders, whose impressive earthen handiwork can be found in that same area.

Armstrong, on the other hand, is not opposed to considering how accurately the unknown Amerindian artists managed to incorporate biblical descriptions of the Devil in their artwork:

"Here do we behold the wings and talons of the eagle, united to the body of the dragon or alligator, with the face of a man, the horns of the black-tailed deer or elk, the nostrils of the hippopotamus, the teeth and beard of the tiger, the ears of the fox, and the tail of the serpent, or fish, with the scales of the salamander, so nicely arranged and fitted together as to preserve the distinctive characteristics of each and produce a picture of all that is the most horrible in animal, fowl, fish, and reptile in a single graphic view. . . . The dragon as above shown is the prototype and representative of Satan, and the serpent is his twin brother, while man is the image of his Maker. . . ."

Some of the Amerindian traditions state that the Piasa was fond of bathing in the Mississippi and was a very rapid swimmer. When it was disporting about in the Father of Waters, it raised such a commotion as to force great waves over the banks.

Dare we conceptualize a surviving Pterodactyl entering the river for either food or fun, raising the river tides as a veritable flying "Nessie," a Loch Ness kind of sea creature with wings?

Other ancient traditions state that when the Piasa was mad—and it seemed as though the sight of an Indian always made it angry—it thrashed the ground with its tail until the whole earth shook and trembled.

The Piasa was generally feared because of its propensity for snatching tribespeople and making off with them. Professor John Russell published an account of the Piasa's insatiable appetite for human flesh in the 1848 July number of *The Evangelical Magazine and Gospel Advocate:*

> He [the Piasa] was artful as he was powerful, and would dart suddenly and unexpectedly upon an Indian, bear him off into one of the caves of the bluff and devour him. Hundreds of warriors attempted for years to destroy him, but without success. Whole villages were nearly depopulated, and consternation spread through all the tribes of the Illini.
>
> A concerned and resourceful chief named Watogo separated himself from his people and fasted and prayed to the Great Spirit for an entire month. On the final night of the fast, the Great Spirit came to Watogo in a dream and instructed him to select twenty of his bravest warriors and to arm each of them with a bow and a poisoned arrow. While the warriors lay in concealment, another tribesman was to stand in open view as bait for the Piasa. The bowmen were to shoot the monster the instant that he pounced upon his prey.
>
> Watogo gave thanks to the Great Spirit and returned to his people to share his vision.
>
> The warriors were quickly selected and placed in ambush as directed. Watogo offered himself as the victim. He was willing to die for his tribe.
>
> Placing himself in open view of the bluff, he soon saw the Piasa perched on the cliff, eyeing his prey. Watogo drew up his manly form . . . began to chant the death song . . .
>
> . . . The Piasa arose into the air and, swift as the thunderbolt, darted down upon the chief. Scarcely had he reached his victim when every bow was sprung and every arrow sent to the feather in his body.

The Piasa uttered a wild, fearful scream, that resounded far over the opposite side of the river, and expired. Watogo was safe. Not an arrow, not even the talons of the bird had touched him. The Master of Life, in admiration of the generous dead, had held over Watogo an invisible shield. In memory of this event, the image of the Piasa was engraved on the face of the bluff.

The above account is filled with the example of self-sacrifice for the good of the people of which so many Amerindian legends are made, but Professor Russell saves the gory documentation for his punch line. He describes how he and a guide managed with great effort to enter the fearsome cave of the awful Piasa.

"The roof of the cavern was vaulted," he writes, "the top of which was hardly less than twenty feet high. The shape of the cave was irregular, but so far as I could judge the bottom would average twenty by thirty feet.

"The floor of the cave throughout its whole extent was one mass of human bones. Skulls and other bones were mingled together in the utmost confusion. To what depth they extended I am unable to decide, but we dug to the depth of three or four feet in every quarter of the cavern and still found only bones. *The remains of thousands must have been deposited there."*

In the legends of the Miami Indians, the Piaza has definite sinister overtones. Once during a fierce battle, the Miamis were pressing their advantage over their traditional enemies, the Mestchegamies, at the upper end of the lower Piaza canyon. As the fighting was reaching its awful climax, the war whoops apparently disturbed the Piasa; and the two fierce, winged creatures emerged from their caves, "uttering bellowings and shrieks, while the flapping of their wings upon the air roared out like so many thunderclaps."

The Piasa swooped low over the heads of the combatants, and each snatched a Miami chieftain in its massive talons.

45

The Miamis became instantly demoralized, believing that the Great Spirit had sent the Piasa to aid and assist their enemies.

On the other side of the coin, the Mestchegamies interpreted the Piasas' choice of Miami chieftains as a sign that the Great Spirit had indeed sent the hideous flying beasts to fight their battles for them. The Mestchegamies slaughtered the demoralized Miami war party and continued their terrible battle thrust into the Miami camp.

The Miamis were so crippled as a nation that the survivors fled toward the Wabash River and did not feel safe until they had crossed its waters. Here they remained for generations before returning to Illinois territory to seek their revenge.

If these stories are true, then the temporary propitious assistance of the Piasa to the Mestchegamies in their desperate battle with the Miamis near Alton, Illinois, proved to be a terrible battle curse instead of a sudden blessing. Soon after the Piasa had flown off with the screaming and struggling Miami chieftains in their talons, the monsters apparently developed a taste for human flesh. Consequently, the Mestchegami came to pay for their victory over the Miamis through an unending sacrifice of their people to feed the ever hungry Piasa, which now seemed insatiable in their forays for manfood.

When did the Piasa conduct their fiendish foraging upon the native tribes? According to Armstrong's little book:

> The time when the Piasa existed in this country, according to the Illini tradition, was "many thousand moons before the arrival of the palefaces," while that of the Miamis says, "several thousand winters before the palefaces came." Though indefinite as to the exact time period, both indicate a very long period of time—many centuries—and may be construed to go away back to the mesozoic or middle-life geological period, known as the age of reptiles, when the monster saurians existed in great numbers and varieties. . . .

Among the most notable of [the giant reptiles] were the pterodactyl, or wing-fingered monstrosity, which in every point of the horrible surpassed the ichthyosaur and plesiosaur. It was an aerial beast, bird, or reptile, with wings shaped like those of the bat. Its bones were hollow like those of the bird, but it had no feathers; and though its bill resembled that of the bittern, it was full of long, sharp teeth like those of the shark. Instead of two legs and feet it had four of each. The fore legs seemed to have come out at the butt of its wings and rested upon them. In shape they resembled human arms, with talons like the eagle . . . but much longer. It probably could walk on its hind legs with folded wings. Its legs, like its arms, were supplied with long and powerful talons. Its spread of wings was from fifteen to twenty-five feet. The fossil remains of some twenty-five species of this monster [*circa* 1887] have been found, and it is sometimes called the pterosaur or flying lizard.

But the most singular monster of the age yet discovered [and its shape and component parts analyzed] is the ramphorhyneus, which seems to be a connective link between birds, beasts, and reptiles. Its body and neck resemble that of the Piasa, while its tail is identical with it, except it is pictured as dragging behind instead of being carried around the body or over its back and head. The shape of the head is drawn to resemble that of a duck, with the long bill of a snipe or bittern, but it is full of sharp, round teeth, like those of the crocodile. It had four legs, with eagle's talons, and a pair of bat-like wings. When on the ground it traveled on all fours, dragging its long tail trailing behind, and when flying it must have wrapped it around its body, under its wings, or around its huge neck. Its entire length from head to tip of tail was probably thirty feet or more.

In many respects the Piasa is a faithful copy of the ramphorhyneus. The form, shape, and description of the Piasa, according to the Indian tradition, were painted from actual sight of the living subject; that of the ramphorhyneus is from collecting its badly decomposed bones, and from their form, shape, and size, constructing an ideal monster.

. . . Thus may the traditions of these Indians be true, and their petroglyphic history of the Piasa may enable the scientist to reconstruct his ramphorhyneus into the shape and form of the Indian's Piasa. If these petroglyphs were the work of the

Indian, and of this we have but little doubt, they show that he had a knowledge, real or traditional, of the existence of these monsters of the geological reptile age.

Our conclusions may be summed up in a few words, as follows:

First. The Indians appeared upon this continent before the extinction of the huge reptiles and saurians of the mesozoic age.

Second. That among the still existing saurians or reptiles when the Indians appeared was one huge monster that could walk, run, fly, and swim, known to the Indians as the Piasa, whose bones have been found and reconstructed into the saurian, or reptile, known to science as the ramphorhyneus.

Third. That this saurian or reptile was of immense size, great strength and voracious appetite . . . and feasted upon Indian flesh.

Fourth. That these petroglyphs were made by the Indians many centuries after the extinction of these monsters as a means of preserving and refreshing their tradition; or, in other words, their tradition was a very old one while these petroglyphs were comparatively of recent date and made by persons who never saw the Piasa, but made them to correspond with the descriptions given in their tradition.

The notion that man or an early species of hominids might have been contemporaneous with dinosaurs has fueled the creative fires of many a fantasy and science fiction writer. Somehow it seems so right that Alley Oop should ride on the back of Dino the dinosaur, and that fur-bedecked "cavemen" should do battle with hungry saurians.

But if man did not exist at the time of the giant reptiles, an alternate scenario maintains that the dinosaurs survived in smaller numbers far later than the seventy million years ago assigned to them by the decrees of paleontologists. Many authors and a few bold-thinking anthropologists have speculated that the legends of dragons, both flying and ground-thundering, might have been the result of man's genetic memory. Again, nearly every romantic tale that concerns itself with lost continents or mysterious islands contains the obligatory sequence of pursuit of the hero or heroine by an angry *tyranosaurus rex* that has not received official word of its extinction.

There hardly seems to be a culture extant that does not have its own accounts of Piasa, dragonlike monsters, swooping down to snatch "man-snacks" in its grasping talons or receiving the tribute of human sacrifice until a tribal hero or knight-errant arrives to slay them. Let us pose the two-pronged question once again: Did a race of men exist during the Age of Reptiles, something like 70 million years ago; or did a certain number of the giant reptiles survive until a few thousand years ago? (We will resist the temptation of asking whether a few might not still be thriving in certain lakes and hidden valleys around the globe.)

Early in January, 1970, newspapers subscribing to the London Express Service carried an item relating the discovery of a set of cave paintings which had been found in the Gorozamzi Hills, twenty-five miles from Salisbury in Rhodesia. According to the news story, the paintings included an accurate representation of a brontosaurus, the 67-foot, 30-ton behemoth that scientists insist became extinct millions of years before man achieved his earthly advent.

Experts agree that the paintings were done by bushmen who ruled Rhodesia from about 1500 B.C. until a few hundred years ago. The experts also agree that the bushmen only painted from life. This belief is borne out by the other Gorozamzi Hills cave paintings, which represent elephants, hippos, deer, and giraffe. According to the news story:

"The brontosaurus, a member of the dinosaur family, can be seen clearly on the rock, its long neck reaching out of a picture of a swamp. . . .

". . . Rhodesian museum authorities refuse to believe that the brontosaurus lived in Rhodesia in recent times. For all the fossilized remains they have examined have been millions of years old.

"Adding to the puzzle of the rock paintings [found by Bevan Parkes, who owns the land on which the caves are located] is a drawing of a dancing bear. As far as scientists know, bears have never lived in Africa."

Getting back to Piasalike, flying reptiles (those monsters would not leave the Amerindians of Illinois alone, either), the November, 1968, issue of *Science Digest* carried the startling thoughts of Mexican archaeologist-journalist José Díaz-Bolio concerning his discovery of an ancient Mayan relief sculpture of a peculiar serpent-bird. The sculpture was found in the ruins of Tajín, located in Totonacapan in the northeastern section of Veracruz, Mexico; and Díaz-Bolio suggested that the serpent-bird was not "merely the product of Mayan flights of fancy, but a realistic representation of an animal that lived during the period of the ancient Mayans—1,000 to 5,000 years ago."

Science Digest observed that a startling evolutionary oddity would have been manifested if such serpent-birds were contemporary with the ancient Mayan culture.

Animals with such characteristics are believed to have disappeared 130 million years ago. The *archaeornis* and *archaeopteryx,* to which the sculpture bears a vague resemblance, were flying reptiles that became extinct during the Mesozoic age of dinosaurs.

And since man did not appear, according to current geological charts, until about one million years ago, there appears to be a 129-million-year discrepancy. The twain (Mayan and serpent-bird) never should have met.

William A. Springstead, writing in *Bible-Science Newsletter*, May 15, 1971, quotes the Bible as an historical source for the coexistence of men and giant reptiles:

In Isaiah 30:6a one reads: "The burden of the beasts of the south: into the land of trouble and anguish, from whence come the young and old lion; the viper and fiery flying serpent." Earlier in 14:29b Isaiah writes: "For out of the serpent's root shall come forth a cockatrice, and his fruit shall be a fiery flying serpent."

. . . The Douay Confraternity translation of Isaiah 30:6 calls the flying serpent the "flying basilisk." The *Jerusalem Bible* refers to the flying serpent of Isaiah 14:19 as a "flying dragon." The New English translation of Isaiah 30:6 speaks of a "venomous flying serpent." Fiery possibly refers to the painfulness induced by the poisonous bite of the serpent.

The importance of reliable historical records to the existence of flying serpents is very great. Although the evidence does not consist of fossil finds, it is evidence nevertheless. Its significance lies in the fact that a bizarre creature long held to be millions of years old is now said to have been on hand only a few thousand years ago. . . .

Dr. Clifford Burdick spent more than thirty years in a study of what appear to be human footprints in strata contemporaneous with dinosaur tracks. Intensive investigation of the several imprints found at Glen Rose, Texas, has convinced Dr. Burdick that these are authentic human footprints.

Now it should be pointed out that Dr. Burdick does have a bias: he is a Creationist, not an Evolutionist. Of course it would have to be conceded that those scientists who ascribe to the theory of Darwinian evolution also have a bias in declaring the footprints to be somehow bogus and unauthentic.

But as a Creationist, Dr. Burdick believes the concept of evolution to be diametrically opposed to the biblical revelations about the true history of man. Creationists interpret the Bible as affirming that in the beginning man was created perfect—physically, mentally, and spiritually. He was also much larger in stature than he is today, even counting our professional basketball and football players. In addition, he was longer lived, attaining an average of over 900 years before the Great Deluge.

Dr. Burdick first began investigating "footprints in stone"

in the early 1950s when the Natural Science Foundation of Los Angeles assigned him to go with four other members to examine the reported man-tracks found in strata contemporaneous with dinosaur prints in and around Glen Rose, Texas. The committee soon learned that men had been cutting dinosaur and human tracks out of the limestone of the Paluxy River bed near Glen Rose since at least 1938. A Mr. A. Berry gave them an affidavit which stated that in September of that year, he and other men found "many dinosaur tracks, several sabre-tooth tiger tracks, and three human tracks" in the river bed.

The committee met Jim Ryals, a man who had dug up and sold dozens of tracks from the Paluxy River area. Many area residents had been selling the tracks during hard times so that tourists' dollars could help them through their economic doldrums. Because of the prints' curiosity value, some less principled individuals had taken to carving the tracks. Ryals told the committee that if the tracks had pressure ridges around the feet, formed by displacement of mud, they were genuine. The committee found that the tracks discovered by Mr. Berry definitely had such pressure ridges.

Dr. Burdick learned that Dr. Roland Bird, field explorer for the American Museum of Natural History of New York City, had also examined the Berry tracks. Describing them in the May 1939 edition of *Natural History* magazine, Bird admitted that he had never seen anything like the tracks, and assessed them as "perfect in every detail." But since the manlike tracks measured 16 inches from toe to heel, Bird declared that they were too large to be human, although the barefoot tracks did show all the toes, insteps, and heels in the proper proportions.

Jim Ryals had accompanied Dr. Bird when he made a special field trip to the Paluxy River to examine the tracks *in situ*. Dr. Bird became less enthusiastic about the prints when he saw them in association with dinosaur tracks, because "man did not live in the age of dinosaurs." So his trip would not be a total waste, Dr. Bird dug up several large Brontosaurus tracks and shipped them to the museum.

In his *Footprints in the Sands of Time,* Dr. Burdick describes the find of Charles Moss of "a sequence of from 15 to 20 perfect giant barefoot human tracks, each about 16 inches in length and eight inches in width. The stride was about six feet until the fellow started to run, when the stride lengthened to nine feet, when only the balls of the feet showed, with the toes. Then the series of tracks disappeared into the bank."

One can visualize a prehistoric giant, caught out in the open by a Brontosaurus or a Tyranosaurus Rex, beating a hasty retreat for a cave . . . or maybe the protection of several of his fellows, who could drive the brutes away with a volley of spears and stones.

The challenge offered by these remarkable footprints in the Paluxy River bed do require a meaningful scientific explanation. Whatever species of creature made these tracks, it was definitely bipedal. The footprints all have about the same length of stride, which would be consistent with a man with a 16-inch foot. The shapes of the prints are more manlike than any other animal known to science.

If the tracks are accepted as being human, then scientists will be forced either to place man back in time to the Cretaceous period or to bring the dinosaurs forward to the Pleistocene or Recent period. While orthodox scientists must undergo a weighty struggle to accept either alternative, Creationists consider the fossil evidence in a much less startling light.

If it could once and for all be established that the fossilized footprints are those of an early race of man, then, according to the Creationists, "the 600-million-year geologic column would suddenly be collapsed to the tune of some 100 million years, the assumed hiatus between the final extinction of the giant reptiles . . . and the advent of man, supposedly some two million years ago.'' While this ''collapse'' would not quite be sufficient to bring geologic time into total harmony with biblical time, Creationists would applaud it as a big step in the right direction.

In referring to the evidence of the Glen Rose tracks, Dr. Burdick states that the general theory of evolution would be dealt a lethal blow, because the geologic record of human footprints contemporaneous with dinosaur tracks ''suggests that simple and complex types of life were coexistent in time past or during geologic ages. . . . This does not harmonize with the hypothesis that complex types of life evolved from lower or more simple forms.

''Evolution implies that through the geologic ages life has not only become more complex, but has increased in size. If evidence from the man-tracks can be used as a criterion, ancient man was much larger than modern man as an average. This harmonized with most fossil life which was larger than its modern counterpart. . . . On the whole, biological life has had to contend with unfavorable environment which has been a factor in its degeneration, rather than its evolution.''

In an earlier work, I wrote at great length about the astonishing find of William Meister, an amateur rockhound, who found what appears to be a fossilized human sandal print with a trilobite, an extinct marine animal, imbedded in the impression made by the heel. Meister discovered the print in July, 1968, while searching for fossils at Antelope Springs, near Delta, Utah. Since the impression was made on what once may have been a sandy beach of the Cambrian period of the Paelozoic Era, the sandal print would have to be an incredible 500 million years old.

Dr. Burdick personally investigated William Meister's find, and while digging in the same area where the rockhound had found the remarkable sandal print, he himself ''was fortunate enough to find on a slab of shale the impression of a child's bare foot with all five toes showing dimly.''

A few days later, Dr. Burdick found a human track "similar to the first one Meister found, evidently made by shoes or moccasins." Professors of the geological department of a leading university conceded that the tracks definitely looked human, but they could not accept their biological origins.

Dr. Burdick's comments regarding the manlike tracks found at Antelope Springs are to be found in his aforementioned "defense statement":

> These tracks with human appearance were preserved in rock hundreds of feet below the present surface of the ground, as though at or near the beginning of some great catastrophic, earth-shaking event that buried many forms of life all together, some marine and some non-marine. This mixing of fossil types is very common all over the world. . . .
>
> If these are verified as human tracks, the discovery will have far-reaching repercussions throughout the scientific world, and especially for stratigraphers and paleontologists. Cambrian fossils such as trilobites, are placed at the bottom of the Paleozoic, some estimated 600 million years before man evolved, according to evolutionary geology. This evidence, if verified, will practically collapse the geologic column. . . .

Frank X. Tolbert has been writing about the alleged mantracks in the Paluxy River for years in his "Tolbert's Texas" column in the *Dallas Morning News*. Tolbert has been consistently skeptical toward any allegations that the prints were made by humans, maintaining that the tracks had been made by giant sloths. But in his January 6, 1973 column, the Dallas journalist reported "what may be the clearest of the so-called 'giant men tracks' yet discovered."

The footprint of "what might have been a huge humanoid" measures 21½ inches in length, 8 inches in width across the front of the foot, and 5½ inches across the instep. Dr. C. N. Dougherty of Glen Rose stated that near the footprint are also the deeply engraved prints of three-toed dinosaurs.

"These men-tracks belong to the Mesozoic Era because the

clearest man-track is exactly eight inches from a trachodon track and on the same layer of rock," said Dr. Dougherty. "The trachodon tracks are as clear and distinct as the man-track."

Columnist Tolbert concedes: "None of the tracks which previous researchers photographed in weeks of work were anything like as humanoid in appearance as the one which Dr. Dougherty found under the waterfall."

According to Dr. Dougherty: "When I discovered this trail of a giant man under the waterfall, I had a feeling that it was one of the most important discoveries since the Dead Sea Scrolls and what is believed to be Noah's Ark on Mount Ararat."

Tolbert explained in his column that it is arduous work seeking out the man-tracks and the dinosaur impressions in the river bed. First of all, one has to wait for the dry seasons when the river had no moving water and the waterfall has gone temporarily dry. Dr. Dougherty admitted that he had swept a big stretch of the river bed during the time of drought before he found "that best of all tracks."

In commenting about the Paluxy River tracks, Tolbert is now prepared to state that the prints "could have been made by very big human beings," who were contemporaneous with the giant reptiles of the Genozoic Era. The thing that most intrigues Tolbert is that the prints are each 21½ inches in length, "the same as the man-like print under the waterfall. And they indicate that these men who were contemporaries of the brontosaurus, if men they were, walked with a stride of 7 feet."

The April 19, 1883 issue of *Nature* carried an account of another unidentified bipedal creature who left humanlike footprints that paced off a remarkable stride. Again, the footprints were found in stratum contemporaneous with prehistoric animals.

The impressions were discovered while building the State Prison near Carson City, Nevada, and were described to the

California Academy of Science by Charles Drayton Gibbs, C.E.:

> These tracks include footprints of the mammoth or some other animal like it, of some smaller quadrupeds, apparently canine and feline, and of numerous birds. Associated with these are repeated tracks of footsteps, which all who have seen are agreed can be the footsteps of no other animal than man. . . . The most remarkable circumstance characterizing them is their great size. In one case there are thirteen footprints measuring 19 inches in length by 8 inches wide at the ball, and 6 inches at the heel. In another case the footprints are 21 inches long by 7 inches wide. There are others of a smaller size, possibly those of women. One track has fourteen footprints 18 inches long. The distance between the footprints constituting a "step" varies from 3 feet 3 inches to 2 feet 3 inches and 2 feet 8 inches, whilst the distance between the consecutive prints of the *same foot* constituting a "pace" varies from 6 feet 6 inches to 4 feet 6 inches. In none of the footprints of the deposit are the toes or claws of animals marked. . . . I need not say that so far as the geological horizon is concerned this discovery does not carry the existence of man beyond the Quaternary Mammalia, with which it has long been pretty clear that he was associated in prehistoric times. Nevertheless it is, if confirmed, a highly remarkable discovery, especially as connected with the curious intimation so concisely made in the Jewish Scriptures, "And there were giants in those days. . . ."

The orthodox consensus concerning the giant manlike footprints near Carson City decreed that they had been made by an extinct species of sloth. Actual skeletons of giant men and women of undetermined origin are less easy to explain away, however.

In the January 11, 1840 issue of the London *Mirror,* a M. le Cat stated that two human skeletons were unearthed near Athens in much earlier times. The length of one was an unbelievable thirty-four feet; the other was an incredible two feet longer. A somewhat shorter relative, only thirty feet in length, was found near Palermo, Sicily in 1548.

Two years after the Sicilian discovery, another skeleton was unearthed—this one measuring thirty-three feet. Finally, another giant skeleton, this one thirty feet long, saw the light of day.

M. le Cat then asserted that in 1705 a twenty-two-foot skeleton was found, and the thigh bone was preserved at Valencia, Spain. Another skeleton was found, the skull of which allegedly held a bushel of corn. If these skeletal remains have indeed been preserved, one wishes that a modern scientist would go to Valencia, or Sicily, and substantiate these old claims.

Other historical accounts of giant-sized skeletons include that given by the Abbe Nazari. This venerable cleric noted that a body exhumed in Calabria, Italy, measured eighteen Roman feet. The average weight of the molars was one ounce.

And Hector Boetius, writing in the seventh book of his classic history of Scotland, declares that the bones of a fourteen-foot man, who had been jokingly nicknamed "Little John," were preserved.

In addition to the giant footprints of manlike bipedal creatures discovered throughout the southwestern United States, other indications that a much larger race had inhabited North America in prehistoric times came with the discovery in Supai Canyon, Arizona, of a petroglyph depicting a mammoth attacking a man. This primitive work of art was found by Harold T. Wilkins, who determined that the beleagured man must have been over ten feet tall, according to the perspective employed by the ancient artist. Amerindians in the vicinity stated that the drawings had been made by the "giants of long ago." One finds that particular phrase cropping up repeatedly in this field of research.

A skull found in one of the many cliff dwellings near Winslow, Arizona, was described by Jesse J. Benton in his *Cow by the Tail,* as being so large that a cowboy's Stetson hat sat on it "like one of those tiny hats merrymakers wear on New Year's." Benton also stated that a gold tooth had been found in the skull, thereby ruling out the likelihood of the head bone having once been connected to a giant animal rather than a giant man.

According to their oral tradition, the Delaware Indians once lived in the western United States. For a long forgotten reason, at some point in their history they migrated eastward as far as the Mississippi River, where they were joined by the Iroquois Confederacy. Both groups of people were seeking land better suited to their rather cultured way of life, and they continued together on their eastward trek.

Scouts sent ahead learned of a nation that inhabited the land east of the Mississippi and who had built strong, walled cities. These people were known as the Talligewi or Allegewi, after whom the Allegheny River and Mountains are named. The Allegewi were considered taller than either the Iroquois or the Delaware, and the scouts saw a good many giants walking among them.

When the two migrating tribes asked permission to pass through the land of the Allegewi, it was denied. Bitter fighting broke out, which continued for a number of years. Eventually, the superior numbers and the determination of the allies prevailed, and the Allegewi fled to the west.

The Allegewi next appear in the legends of the Sioux, whose tradition tells of a confrontation with a race of "great stature, but very cowardly." The Sioux, who were surely among the ablest warriors and the most resourceful fighting men of any of the Amerindian tribes, exterminated the Allegewi when the giants sought to settle in what is now Minnesota.

Is there any archaeological evidence to support these Indian legends and traditions?

Rising out of the earth in Ohio, Minnesota, Iowa, and other

states are the huge Earthworks of the mysterious "mound-builders." The mounds scattered throughout the Midwest were apparently raised by the same unknown people, and the earthworks are extremely large.

Do giant mounds indicate giant people?

Enormous weapons, including a copper ax weighing 38 pounds, have been found in these mounds. It is difficult to imagine the average-sized Amerindian, as we first know him at the time of the European invasion, casually wielding a 38-pound ax.

However, outsized weapons and implements alone are not proof of a giant race, and neither are huge monuments. The former can be works of art, the latter could be objects of religious commitment. The best proof of a race of giants in North America—or anywhere else—would be the discovery of the skeletons of these people.

Such skeletal proof may have been found in Minnesota.

Several years ago, two brothers living in Dresbach, Minnesota, decided to enlarge their brick business. To do so, they were forced to remove a number of large Indian mounds. In one of the huge earthenworks they removed the bones of "men over eight feet tall." Unfortunately, for purposes of settling a centuries-old debate concerning "giants in the earth," these bones crumbled when exposed to the air. Their existence rests on the word of the Minnesotans who witnessed the excavation.

In La Crescent, Minnesota, not far from Dresbach, mound-diggers reportedly found large skillets and "bones of men of huge stature." Over in Chatfield, mounds were excavated, revealing six skeletons of enormous size. Unusually large skeletons of seven people buried head down were discovered in Clearwater. The skulls in the latter find were said to have had receding foreheads, and teeth that were double all the way around.

Other discoveries in Minnesota included "men of more than ordinary size" in Moose Island Lake; several skeletons, one of "gigantic size"in Pine City; ten skeletons "of both sexes and of gigantic size" in Warren (buried with these particular specimens were horses, badgers, and dogs).

Could these huge skeletons of gigantic "indians" have once supported the flesh and internal life-support systems of the last of a proud prehistoric race who defied the monster reptiles and built an extensive empire of walled cities throughout the world? In historic times, their numbers severely reduced, their great cities all but destroyed, they may have seemed cowardly to more aggressive tribes such as the Sioux.

Throughout the Old Testament, the wandering Israelites encountered many gigantic peoples, whom they slew upon "guidance" from the Lord—and the strong sword arms of such warriors as Joshua and David. In early struggles for "lands of milk and honey," the nomadic tribes of the Middle East, Central Europe, and the Americas might very well have massacred the last of our giant species' ancestors.

The New York Times on December 2, 1930, carried an item that told of the discovery of the remains of an apparent race of giants who once lived at Sayopa, Sonora, a mining town 300 miles south of the Mexican border. A mining engineer, J. E. Coker, said that laborers clearing ranchland near the Yazui River "dug into an old cemetery where bodies of men, averaging eight feet in height, were found buried tier by tier. . . ."

On February 14, 1936, *The New York Times* ran a piece datelined Managua, Nicaragua, which stated that the skeleton of a gigantic man, with the head missing, had been unearthed at El Boquin, on the Mico River, in the Chontales district. "The ribs are a yard long and four inches wide and the shin bone is too heavy for one man to carry. 'Chontales' is an Indian word, meaning 'wild man.' ' "

The year 1936 must have been a good one in which to find gigantic skeletons. In its June 9 issue, *The New York Times* published this item with a Miami, Florida dateline:

A tale of human skeletons eight feet long imbedded in the sand of an uninhabited little island off Southern Florida was brought here today by three fishermen. They exhibited a piece of one skull containing six teeth.

E. M. Miller, zoologist at the University of Miami, said the mandible was that of a man and was probably several hundred years old. "It is entirely probable that this find might be important," he commented. The men said that the skulls were unusually thick, the jaws protruded, and the eye sockets were high in the head.

In his book *Forbidden Land*, Robert R. Lyman wrote of an unknown tribe of American Indian giants who had the added distinction of *horns* growing from their heads:

At Tioga Point . . . a short distance from Sayre, in Bradford County [Pennsylvania], an amazing discovery was made. Dr. G. P. Donehoo, State Historian and a former minister of the Presbyterian Church in Coundersport, together with Prof. A. B. Skinner of the American Investigating Museum, and Prof. W. K. Morehead of Phillips Andover Academy, uncovered an Indian mound. They found the bones of 68 men which were believed to have been buried about the year 1200. The average height of these men was seven feet, while many were much taller. On some of the skulls, two inches above the perfectly formed forehead, were protuberances of bone, evidently horns that had been there since birth. Some of the specimens were sent to the American Investigating Museum.

. . . In December 1886, W. H. Scoville of Andrews Settlement discovered an Indian mound at Ellisburg. When opened, the skeleton of a man was found. It was close to eight feet in length. Trees on and around the mound indicated that burial had been made at least 200 years before.

The notion of seven-foot giants with horns on their heads

may immediately conjure up visions of "devils," "fallen angels," "gargoyles," and other monsters that seem to have won permanent positions in man's collective unconscious and his genetic memory. Just as some theorists conjecture that the universal legend of the dragon is mankind's inherited memory of the dinosaur, so do others muse that the ubiquitous giants and grotesque monsters of our folklore cloak the genetically transmitted memory of former races of beings that were somehow more than men—or something other than man as we know him.

A medical doctor specializing in circulatory ailments by day and conducting strange archaeological digs by night, Dr. Javier Cabrera of Ica, Peru, has collected more than 15,000 engraved stones that appear to depict a very special race of man living side by side with the great prehistoric reptiles. Biologist Ryan Drum, writing in the May, 1976 issue of *INFO, the Journal of the International Fortean Organization,* told of his visit to Dr. Cabrera and of his own examination of the strange petroglyphs.

The rocks, Drum states, vary in size from fist-sized cobbles to boulders, but they were all covered with the weird petroglyphs.

"Many of the largest," according to Drum, "depicted strange five-fingered people fighting or at least interacting with what looked like dinosaurs. . . . What bizarre people, with pointy tongues and noses which began in the middle of their foreheads, reminiscent of the Mayas, who allegedly used chicle putty and clay to extend their nose as if it originated from the center of their forehead. . . . The fingers are similar to the skinny digits clasped around the pouchy stomachs of Easter Island Statues. The people were simply clothed about the loins, sported Egyptian-like headgear, have battle-axes like Vikings, and seem disproportionate to the 'dinosaurs.' "

Dr. Cabrera claimed that he began finding the mysterious

stones near Ica more than twenty years ago after an earthquake prompted a landslide which exposed a large deposit of the picture rocks. Many of his most severe detractors claim that Cabrera hires Indians to carve the petroglyphs; and to the frustration of his defenders, the doctor refuses to divulge the exact location of the mysterious cache of stones so that everyone might examine the site for himself. It is Dr. Cabrera's theory that the petroglyphs were fashioned by protopeople who lived near Ica from 250,000 to one million years ago.

Ryan Drum describes the most astonishing stones, which depict the five-fingered, thumbless protopeople performing wonders of medical science: "Poor lighting and shadows could not hide astonishing petroglyphs depicting brain surgery, heart transplants, lung, kidney, and liver operations and transplants . . . and strange ceremonies of no known or named counterpart.

". . . Here were operating tables . . . surgical knives, local and general anesthesia, sutures, and more. . . . The figures were crudely drawn but the organs were masterpieces. Gross hatching technique was used. The style seemed consistent, but a peculiar mixture of care and haste."

Dr. Cabrera believes that the rocks constitute "an ancient library . . . scratched in plastic mud and hardened over the millenia." Ryan Drum managed to bring two of the stones back with him:

> I have examined the rocks at 30 and 60 magnifications in a stereo microscope to study the grooves, and found no obvious grinding or polishing marks or any other evidence of rotary power tool use in making the fine regular grooves. I am not sure how to date the rocks, since they are susceptible to potassium-argon dating only if they are in a volcanic deposit. If Cabrera is right and the rocks are genuine as claimed, they are incredibly valuable and should not only be held in awe, but studied thoroughly as products of human intelligence. If they are a hoax, their existence, number, detail, and bulk represent

an enormous input of human resources and should be regarded as a fine combination of intellect and imagination. . . .

The *Peruvian Times*, August 25, 1972, sent their own reporters to investigate the stories of the weird rocks that showed things such as men flying on pterodactyli, bizarre animals, and open-heart surgery—all being conducted by a lost race of men who allegedly lived more than 70 million years ago.

According to the *Peruvian Times*, the dinosaur stones are the simplest pieces in Dr. Cabrera's collection. Larger stones portray the protopeople's society, its mythology, and astronomy, including a calendar system of thirteen months, each of 28 days.

"Related stones also depict the horse and, symbol of our progress, the wheel," stated the *Peruvian Times*. " 'These,' explains the doctor confidently, 'were attributed to the exclusive use of the gods, so they were never incorporated into the practical development of the race.

" 'When the Conquistadores first appeared in South America . . . they were greeted as gods because they rode horses. Yet horses, we are told, were unknown before the Spaniards came. The theory is simple. You may fear something because it is unknown, but worship requires recognition.' "

The petroglyphs portray the protopeople as possessing no opposing thumbs. Because even the most primitive apes have opposing thumbs—both in the fossil record and in the world today—Dr. Cabrera takes this fact as additional proof that his protopeople predated the apes. Dr. Cabrera believes that the prehistoric men were destroyed by a cataclysmic disaster after having accomplished many important contributions to posterity, including the pyramids of Egypt.

"In their particular way, [the Cabrera rocks] definitely present man alongside the dinosaur," concludes the *Peruvian Times*. "They give very clear pictures of the operations which

20th century surgeons are only just contemplating. So that even if you think the engravings are only two thousand or two hundred years old, or even twenty or two years old, they are still remarkable, and you still have to account for their incredible pictures.''

Dr. Cabrera has said that when he receives government guarantees that his studies and his stones will be protected, he will at that time reveal the source of his incredible prehistoric "library."

If the Cabrera stones should be authenticated beyond all doubt and if the remarkable fossilized footprints discussed in this chapter should be accepted as genuine, then all our accepted theories about man's genesis will be turned inside out. Our image of man's evolutionary history will be desperately in need of refocusing. Man, according to the evidence presented in this chapter, may not only have lived tens of millions of years earlier than our orthodoxies have ever permitted, but a strange, five-fingered species with pointy tongues and noses may have thrived at the time of the dinosaurs.

Giant men and giant reptiles?

The twain not only might have met, they may have interacted for more than a million years.

CHAPTER FIVE

Magnificent Manna Machines, Ancient Aviators,

and Other Wonders of Prehistoric Science

In May, 1976, scientists became very excited by the discovery in Thailand of an ancient Bronze Age culture that might represent the oldest evidence of sophisticated technology in man's history. The site at Ban Chiang, according to the archaeologists, was flourishing when the cultures that later became ancient Egypt and, still later, Greece, were just beginning.

The forgotten people of Ban Chiang, in northeast Thailand near Laos, may have been farming rice and keeping draft animals as early as 5000 B.C. They were using metal alloys as early as 3600 B.C.—and possibly 1,500 years before that.

According to Dr. Froelich Rainey, director of the University of Pennsylvania museum, the discovery at Ban Chiang "challenges all the assumptions that have long been held about the development of our modern cultures."

Our scientists are continually being forced to push back the date of man's technological origins.

In July, 1935, *Scientific American* carried the news of "surprising discoveries'' made in the excavation of Tell Asmar, site of the ancient city of Eshnunna, 50 miles from Baghdad. At that site, clear glass of a bluish color was found dating back to 2700 B.C. and a dagger handle with iron rust, dating to the same time, were found. Both objects appeared about 1,500 years earlier than previously expected, thereby suggesting a technology more than 4,500 years old.

In addition, Dr. Henri Frankfort of the Oriental Institute of the University of Chicago reported "the discovery of a private house with four arched doorways, three of which were perfectly intact."

Dr. Frankfort said that it was new knowledge "that the arch was used at all in this early period. Its discovery was as unexpected as that of the window [which was also found at this site for the first time in the history of Babylonian excavation]."

In one of the temples unearthed, the archaeologists were impressed with "the elaborate arrangements for sanitation." They disvovered no less than six toilets and five bathrooms.

The plumbing equipment was "connected to drains which discharged into a main sewer, 1 meter high and 50 meters long. This was vaulted over with baked brick and ran along the outer wall of the building beneath the pavement of a passageway."

As the excavators were tracing one drain, they came upon a line of earthenware pipes: "One end of each section was about eight inches in diameter while the other end was reduced to seven inches, so that the pipes could be coupled into each other just as is done with drain pipes in the 20th Century A.D.''

According to the July 27, 1959 issue of *Chemical Engineering*, the ancient Romans also built their pipes, valves, and fittings according to modern standards. At the Fifth World Petroleum Congress, Mario Fera, senior engineer for the Italian firm Compagnia Tecnica Industrie Petroli, S.P.A., of Rome, exhibited an 80-pound valve that had been salvaged from one of Emperor Caligula's yachts. The valve was made of a zinc-free, lead-rich, anticorrosion, antifriction tin bronze. Quoting from *Chemical Engineering*:

> The Caligula valve was found submerged at the bottom of Lake Nemi in Rome. Although 19 centuries old, it still exhibits highly polished surfaces and retains its plug tightly.
> Other valves found in and about Rome and Naples are equally well preserved, Dr. Fera said. One found at a reservoir in a Roman villa was used regularly until three years ago by the peasants whenever they required water for irrigation. In 1956 it was decided that the cost of the valve had been amortized, and it was replaced after 20 centuries of service.

Somehow we nod our heads in understanding and in respect for the genius of the early people of the Mediterranean coastlines. After all, the seeds of Western civilization and culture, which culminated in the magnificence that is our very selves, were sown in that area. Even we laymen become somewhat uncomfortable when we learn about such sites as Ben Chiang in Thailand, because we are forced to deal with the reality that there appear to have been several very ancient "cradles of civilization."

In 1968, Dr. Korium Megertchian, a Soviet archaeologist, unearthed what is as of this writing the oldest large-scale metallurgical factory in the world at Medzamor in Russian Armenia.

Incredible as it may seem, at this site, 4,500 years ago, an unknown prehistoric people worked more than 200 furnaces in order to produce an assortment of vases, knives, spearheads, rings, bracelets, and other metal items. Safety-conscious, the prehistoric craftsmen of Medzamor wore mouth-filters and gloves while they worked. Evidence present at the site indicated that they made their wares of copper, lead, zinc, iron, gold, tin, manganese, and as many as fourteen varieties of bronze.

The productive smelters also produced an assortment of metallic paints, ceramics, and glass, but as the scientists admitted freely, the most anachronistic item was several pairs of tweezers, fashioned of steel and unearthed from layers dating back before the first millenium B.C. Metallurgical experts in the Soviet Union, the United States, Britain, France, and Germany later verified the claim that the steel used in the tweezers was of an exceptionally high grade.

Reporting in *Science et Vie*, July, 1969, French journalist Jean Vidal stated his belief that such finds as those at Medzamor indicate an unknown period of technological development.

In Vidal's opinion: "Medzamor was founded by wise men of earlier civilizations. They possessed knowledge they had acquired during a remote age unknown to us, that deserves to be called scientific and industrial."

In his privately published manuscript, *The Legacy of Methuselah*, Joey Jochmans makes the observation that what makes the Medzamor metallurgical site especially interesting to those in the Judeo-Christian tradition is that "it is within fifteen miles of Mount Ararat—the landing site of the survivors [Noah and his family] of the destroyed Antediluvian civilization."

According to what we know of aluminum, it was not discovered until 1807 and it was not produced successfully in pure form until 1857. Even today the process of extracting aluminum from bauxite mineral is a very complicated process, which involves the utilization of a Reverbier oven, a refraction chamber and regenerator, as well as electrolysis and the production of temperatures which must exceed 1,000 degrees centigrade.

With that bit of technical trivia out of the way, we may now deal with the puzzling discovery that was made in China at the burial site of Chou Chu, a general of the Tsin era, who lived about 265 to 316 A.D. The astonishing artifact in this case was a belt fastener that was not only made of metal with openwork ornamentation, but was composed of an alloy of 5 percent manganese, 10 percent copper, and *85 percent aluminum.*

Although orthodox researchers are able to concede that the exquisite glass miniatures produced by Egyptian craftsmen as early as 3000 B.C. are "among the most astounding achievements in the history of glassmaking," they believe that "industrial chemistry in antiquity, though tremendously impressive, was strictly a trial-and-error affair."

Writing in the September, 1964 issue of *National Geographic*, Ray Winfield Smith states: "The ancients of the Mediterranean and the East were intelligent and practical people. Yet when I speak of cerium and lanthanum, of antimony and manganese, of lead oxides, soda, and lime, you must not think that the artisans of those times understood the chemistry of these ingredients as we do. . . . Glassmakers simply knew that certain substances in nature—in rock, sand, earth, or ashes—gave special properties to their products. . . ."

If glassmaking was such a hit-and-miss, trial-and-error procedure, we must certainly wonder how such a massive piece of glass as that found in a cave near Haifa, Israel in 1966 ever came into being. Although this area has been a famous glass-making region since the time of the Phoenicians, this slab of glass is eleven feet long, seven feet wide, 1½ feet thick, and weighs 8.8 tons. The piece of glass is a solid one, raspberry-colored with greenish streaks.

Commenting on the anomalous glass find in the Autumn, 1972, issue of the *INFO Journal*, Ronald J. Willis journalistically contemplated the fact that there are only two masses of glass larger than the aforementioned Haifa-piece known to man:

"Both are the casts for the huge mirrors of the Mount Palomar telescopes. The archaeologists date the slab at 1,000 to 1,400 years old. It conceivably could be the remains of a great melt made in a huge furnace that was allowed to cool for some reason. Perhaps it was meant to be broken up into smaller chunks to be remelted and shaped into bottles and other glass objects. But how did these ancient people develop the enormous amount of heat necessary to melt the ingredients into this enormous mass of glass?''

Some theorists might be bold enough to state that the ancients were using some of the glass to fashion light bulbs for their electrical systems!

Many readers will be familiar with the discovery of a number of ancient dry-cell electric batteries made by Dr. Wilhelm Konig in 1938. The find was realized in a number of clay pots two thousand years old which had been unearthed at Kujut Rabua, a village to the southeast of Baghdad. It was while he was working for the State Museum in Baghdad, Iraq, that Konig, a German archaeologist, was bold enough to demonstrate that the ancients utilized electricity.

According to the reports of Konig, the clay pots were each six inches high, and each contained a cylinder of sheet copper five inches high and one and one-half inches in diameter. The edges of the copper cylinders were soldered with a 60-40 lead-tin alloy, which is comparable to the best solder we have today. The bottoms of the cylinders were capped with copper discs and sealed with bitumen or asphalt. An additional insulating layer of bitumen sealed the tops of the pots and held in place iron rods which were suspended into the center of the copper cylinders and bore evidence of having been corroded by an acid solution.

Confirmation of Dr. Konig's assertion that the clay pots with their configurations of copper, iron, and acid constituted ancient electric dry-cell batteries came when science historian Willey Ley, in cooperation with William F. M. Grey of the General Electric High Voltage Laboratory in Pittsfield, Massachusetts, structured a duplicate model of the clay-pot batteries. By adding copper sulfate, acetic acid, or citric acid (all of which were known and in common usage two thousand years ago), they found that their replica of the Babylonia battery was capable of producing between one and one-half and two volts of electricity.

We know that nature has never been without its spontaneous displays of the might and power of electrical manifestations, but the "control" of electricity has always seemed a part of the modern era. In point of fact, generation of electrical current by the means of dry-cell electric batteries was not invented in Europe until 1800. To consider that gold- and silversmiths were using electricity to assist them in the plating of metal objects more than two thousand years ago is to undergo a bit of historical disorientation.

Other batteries have since been located. In some sites, groups of clay pots have been found together with thin iron and copper rods, which may have been used to connect the pots into a series so that stronger voltage might be produced.

The vast majority of the ancient batteries that have been located date from the Parthian period of Persian occupation of the region, between 250 B.C. and 650 A.D. But electroplated objects—or so they certainly appear—have been discovered in Babylonian ruins that date back to 2000 B.C.

Auguste Mariette, a famous nineteenth-century French archaeologist, unearthed a number of electroplated artifacts at a depth of 60 feet in the area of the Sphinx of Gizeh. In the *Grand Dictionaire Universel du 19ieme Siècle*, Mariette described the artifacts as being "pieces of gold jewelry whose thinness and lightness makes one believe they had been produced by electroplating, an industrial technique that we have been using for only two or three years."

Joey R. Jochmans is among those who have speculated that since the ancient Egyptians possessed electricity to electroplate gold jewelry, they might also have utilized it to illuminate the intricately devised passageways in their tombs.

Jochmans points out that the traditionally accepted means of illumination available to the Egyptians would limit their lighting to torches and oil lamps, but no trace of smoke or soot has been found on the ceilings of the pyramids or the subterranean tombs of the Pharoahs in the Valley of the Kings.

"It has been thought that perhaps the Egyptians used some complicated system of lenses and mirrors to bring sunlight into the burial chambers," Jochmans says, "but no remains of any such system have ever been found. There are a number of ancient tombs with tunnels and passageways that are too complex for a mirror system to have brought sufficient light into the inner chamber. The only other alternative is that the Egyptians had a smokeless light source."

But could that "smokeless light source" really have been electricity?

Some theorists, Jochmans among them, have even suggested that the wall engravings in the Temple of Dendera, built during the Ptolemaic period, depict men handling what could be Crookes tubes, the forerunner of the modern television tube. Jochmans writes:

> When the [Crookes] tube is in operation, the ray originates where the cathode electrical wire enters the tube, and from here the ray extends through the length of the tube to the opposite end. In the temple picture, the electron beam is represented as an outstretched serpent. The tail of the serpent begins where a cable from the energy box enters the tube, and the serpent's head touches the opposite end. In Egyptian art, the serpent was the symbol of divine energy.
>
> . . . The Temple picture shows one tube, on the extreme left of the picture, to be operating under normal conditions. But with the second tube, situated closest to the energy box on the right, an interesting experiment has been portrayed. Michael R. Freedman, an electric and electromagnetic engineer, believes that the solar disc on Horus' head is a Van de Graaff generator, an apparatus which collects static electricity. A baboon is portrayed holding a metal knife between Van de Graaff-solar disc and the second tube. Under actual conditions, the static charge built up on the knife from the generator would cause the

electron beam inside the Crookes tube to be diverted from the normal path, because the negative knife and negative beam would repel each other. In the Temple picture, the serpent's head in the second tube is turned away from the end of the tube, repulsed by the knife in the baboon's hand.

Such theorists as Jochmans are convinced that every aspect of the Temple picture represents an important feature of a "serious scientific experiment."

With the artifacts that continue to be unearthed at important dig sites around the world, one cannot state with any pronounced degree of certainty that the ancient Egyptians were not in possession of electron tubes. But it seems rather likely that if they were manipulating such electronic objects, they may have "inherited" them from a culture superior to their own.

And, of course, that suggestion provokes the controversy about "ancient astronauts" from some extraterrestrial source and revives the much older conflict concerning superscientists from a lost continent, such as Atlantis. Indeed, if one speculates about "worlds before our own," then one must be prepared to be open to either hypothesis.

At this moment in time and space there may not be enough dramatic evidence to convince the academics that either hypothesis is a tenable one—that any spacecraft from another planet landed here to take possession of our liquid green planet or that a mighty nation sank under the Atlantic Ocean—but the evidence seems to be steadily accumulating that civilization has been cyclical upon this terrestrial globe and that as rabbinical literature states, "worlds upon worlds there were before Adam was."

In regard to the ancient Egyptian electron tubes, electromagnetics engineer Professor S. R. Harris identified a box-and-braided cable in the picture as "virtually an exact copy of engineering illustrations used today for representing a bundle of conducting electrical wires." The cable runs from the box the full length of the floor and terminates at both the ends and at the bases of two peculiar objects resting on two pillars. Professor Harris is said to have identified these representations as high voltage insulators.

I have really never understood how anyone, no matter how expert, could positively identify highly technical artifacts contemporary with our culture solely on the basis of an ancient Egyptian's artistic impressions of a ritual or a workday chore contemporary with his own culture. I wonder if such expert witnesses can really positively identify such things as electron tubes and rocket sleds in the artworks of the ancients or if they are simply speculating that the old paintings or sculptures appear to represent such modern artifacts: "Yes, by thunder, that certainly does look like a transistor radio in that old Pharaoh's hand!" But the expert witness knows full well that the ancient priest or prince must really be holding an instrument associated with some long forgotten ritual of court or clergy.

On the other hand, I would like to think that I could recognize the drawing of a typewriter if I saw it etched on some Cro-Magnon's cave.

Within the last decade or so, some very fine minds—both in and out of the ecclesiastical establishment—have begun to theorize quite openly about the possibility of the representatives of some advanced technology having interacted with the prophets and revelators of the Bible. Some have written serious works about UFOs and Holy Scriptures. Others have entertained the notion of such activity from church pulpit or classroom lectern.

As an example pertinent to this chapter, some of these speculators have wondered if Noah might not have possessed electricity on board his gigantic ark. The point of discussion has to do with the fact that there are two quite different Hebrew words used to describe what is translated as a "window" or "opening" in the ark.

There is *challon* or "opening" out of which Noah released the birds that would bring back word of the Great Deluge's having abated. But the first reference, in Genesis 6:16, employs the word *tsohar*, which does not mean either window or opening, but is a Hebrew word so old that most scholars are uncertain of its exact interpretation. According to Jochmans:

> Where it [*tsohar*] is used on twenty-three other occasions in the Old Testament, it is given the meaning, "a brightness, brilliance, the light of the noon-day sun." Its cognates have the word refer to something that "glistens, glitters or shines." Many Jewish scholars of the traditional school identify the *tsohar* as "a light which has its origins in a shining crystal." Hebrew tradition for centuries has described the *tsohar* as an enormous gem or pearl that Noah hung up from the roof of the Ark, and by power contained within itself lighted the entire vessel for the duration of the Flood voyage.

Atlantean enthusiasts will surely seize upon the above reference to a power source contained within a shining crystal as suggestive of the legendary crystals said to have served as everything from energy plants to destructive weapons for the lost science of Atlantis. Perhaps, such an advocate of Atlantean lore will theorize, the story of the Great Flood is symbolical of the sinking of Atlantis. Noah may have been a surviving Atlantean piloting his mercy ship by means of one of the power crystals.

Some several generations after Noah and his family disembarked on Mount Ararat and returned to repopulate a purged world, the great Moses led his enslaved tribesmen away from a civilization that had once again become corrupted so that they might search for a Promised Land.

While the Children of Israel wandered through wilderness and desert, they were kept alive by eating manna, a mysterious foodstuff that appeared every morning. Now two scientists have theorized that rather than having come from heaven, the food may have been a single-cell protein manufactured in a special fermentation unit.

According to Exodus 16:14, manna appeared as a "small round thing, as small as the hoar frost on the ground." Moses told the puzzled Israelites that they were seeing the "bread" which the Lord had promised them, and he prescribed a ration for the people dependent upon family size.

The taste of manna is described in Exodus 16:31 as being like "coriander seed, white; and the taste of it was like wafers made with honey." As a nutritious food, manna must have been adequate, for the children of Israel "did eat manna forty years, until they came to a land inhabited [Canaan]" (16:35).

In the April 1, 1976 issue of *New Scientist*, George Sassoon, a linguist and electronics consultant, and Rodney Dale, a biologist and freelance engineering writer, set forth their hypothesis that rather than having been produced by heavenly manifestation, the secretion of parasites on tamarish trees, or the generally available lichen *Lecanora esculenta*, manna may have been manufactured by a "machine from some forgotten people." Sassoon and Dale believe that they found the answer to the enigma of manna in the *Kabbalah*, the esoteric teachings of Judaism.

In the *Kabbalah*, the two scientists found a detailed physical description of a god named the "Ancient of Days." The androgynous god consisted of a male part and a female part, which is quite typical of mystery religions. What struck Sassoon and Dale as truly odd about the god was the fact that it appeared to be a god that could be taken to pieces and reassembled.

The minute details of this procedure caused the two scientists to "look closely at the texts, uncluttered by peripheral verbiage, and to decide that there was a high probability that they describe not a god but a machine in anthropomorphic terms." (Nontechnological people might, for example, refer to an automobile's headlights as its "eyes," its wheels as its "legs," its exhaust as its "breath.")

When Sassoon and Dale read in one of the texts such passages as the following, they believed the words to be describing a machine for making manna: "Into the skull . . . distils the dew from the white head . . . and from this dew they grind the manna . . . and the manna did not appear to be derived from this dew except at one time: the time when Israel was wandering in the desert."

The two scholars found the most substantial verses to be 51-73 of the book called HADRA AVTA QDISHA (Lesser Holy Assembly) from *Kabbalah Denudata* (1644):

. . . The top skull is white. In it there is no beginning or end. The hollow thing of its juices is extended and is made to flow. . . . From this hollow thing for juice of the white skull the dew falls every day into the small-faced one. . . . And his head is filled, and from the small-faced one it falls. . . . The Ancient Holy One is secret and hidden. And the upper wisdom is concealed in the skull which is found [i.e. can be seen] and from this into that the Ancient One is not opened [i.e. there is no passage visible]. And the head is not single because it is the top of the whole head. The upper wisdom is inside the head: it is concealed and is called the upper brain, the concealed brain, the brain that appeases and is quiet. And there is no [man] that knows it. . . .

Three heads are hollowed out: this inside that and this above the other. One head is wisdom; it is concealed from that which is covered. This wisdom is concealed; it is the top of all . . . [heads of the other wisdoms. The upper head is the Ancient and Holy One, the most concealed of all concealed ones. It is the top of the whole head, the head which is not a head [not an ordinary one], and is not known. And because of this, the Ancient Holy One is called "nothing." And all those hairs and all those cords from the brain are concealed, are all smooth in the carrier. And all of the neck is not seen.

. . . There is one path that flows in the division of the hairs from the brain . . .

Sassoon and Dale feel certain that other parts of the esoteric texts indicate that the "hairs" and "cords" are wires
and pipes. "Wisdom," they conclude, is a liquid utilized in
the process of manufacturing manna. By retranslating the
original Aramaic and by reflecting upon all the possible mean-
ings where they felt there was a choice, the scientists postu-
lated the manna machine as follows:

> At the top is a dew-still: a refrigerated, corrugated surface
> over which air is drawn, from which water condenses. This is
> fed to a container in the center of which is a powerful light-
> source for irradiating a culture, possibly of Chlorella-type al-
> gae. There are dozens of strains of Chlorella, and the balance
> of protein, carbohydrate, and fat in a chosen strain can be var-
> ied by choosing the appropriate conditions of growth for the
> culture.
> This algal culture circulates through pipes which permit an
> exchange of oxygen and carbon dioxide with the atmosphere,
> and also dissipate heat. The *Chorella* sludge is drawn off into
> another vessel where it is treated so that the starch is partially
> hydrolysed to maltose, which is then burnt slightly to give the
> honey-and-wafers flavor. . . . The dried material is then fed
> to two vessels. One is emptied daily to provide the day's sup-
> ply, and the other fills slowly during the week so that two days'
> supply is available on the eve of the Sabbath. . . .

Sassoon and Dale feel that the account in Exodus which describes the Israelites gathering the manna from the ground
each morning is totally unacceptable. Since the Ancient of
Days had to provide an "omer" of manna per day per family
for 600 families, the two researchers computed the output to
be the equivalent of about 1.5 cubic meters of manna per day.
Once the Israelites had ceased their forty years of wandering and the manna machine was no longer needed, the question arises as to what happened to the Ancient of Days.
In Joshua 5:12, we learn that the "machine" stopped working: "And the manna ceased on the morrow after they had
eaten of the old corn of the land; neither had the children of

Israel manna any more; but they did eat of the fruit of the land of Canaan that year.''

When Jericho fell to the Israelites, the manna machine, now revered as a holy object, was kept at Shilon. The Philistines, perpetual nemeses of the Israelites, once stole the machine, but they hastily returned it when their usage of it afflicted them with "emerods [tumors] in their secret parts" (I Samuel 5:8-12).

If we assume with Sassoon and Dale that the Ark of the Covenant included the manna machine, then King David reinstated it as a ritual object in a tent at Jerusalem, and his son Solomon constructed the first temple to house it. When the temple was sacked by Nebuchadnezzar in 586 B.C. the manna machine was destroyed.

Sassoon and Dale speculate that such machines as the Ancient of Days would be "essential equipment on board space vehicles," in that they could perform the dual functions of manufacturing food to eat and expelling oxygen to breathe. The two scientists remind us that the Soviets have constructed such a machine and utilized it to purify the air on board a Salyut space laboratory for some months. The algal cultures were fertilized with the cosmonauts' own excreta and possibly for that reason the "manna" was not eaten.

It would seem, then, that our present fermentation technology is not quite as advanced as the science responsible for the Ancient of Days. According to Sassoon and Dale, the principal component missing is the "high intensity, high efficiency light source," but they concede that "laser optics might just about meet the requirement."

Where did the Israelites acquire their magnificent manna machine?

Sassoon and Dale admit that it is tempting to speculate that the Earth was visited by intelligences from an extraterrestrial point of origin some 3,000 years ago, who, for some reason, looked with favor upon Moses and his wandering tribe and provided them with a machine that could feed them with the inexhaustible persistence and regularity of a Jewish mother. Such a theory, the two scientists decided, "raises as many problems as it solves, and we would prefer not to propound such a hypothesis today."

The question of ancient aviators is another area that raises as many problems as it solves, and we shall next deal with that intriguing subject.

A great airship constructed about 1279 A.D. by Ko-Shau-King, Chief Astronomer to Kublai, was used at the coronation of the Emperor Eo-Kien in 1306. Marco Polo records that he saw the Great Armillary Sphere at the Court of Cathay, and Father Vasson, a French missionary in Canton, states that he saw an account of the airship recorded in a letter dated September 5, 1694.

According to F. T. Miller's *The World in the Air*, the Chinese claimed to have had "a system of signals by which 'different-toned trumpets sounded from the tops of high hills and gave notice of impending changes of wind and weather, for use by navigators of dirigible balloons.'"

Many of those men and women who theorize about the enigmatic drawings of gigantic size that have been carved into the Nazca desert, and especially *Chariots of the Gods?* author Erich von Däniken, believe that the lines were actually used as landing strips for the craft of ancient astronauts. But recently such explorers as Bill Spohrer and Jim Woodman have asserted their contention that the lines were actually guiding strips for balloonists. In their opinion, the "gods" from outer space theory is "hogwash." The ancient Peruvians had only to observe smoke rise to apply the "warm air rises" principle to ballooning.

Spohrer and Woodman state that there are many Peruvian legends telling of men who flew.

Supportive data was found in the records of a Portuguese Jesuit Priest named Bartolomei De Gusmao, who, in 1690, returned from the Peruvian jungles claiming that he had witnessed Peruvians flying in balloons. Apparently De Gusmao substantiated his claim by flying a model for the King of Portugal in 1709. Since the flight was officially recorded, De Gusmao was aloft eighty years before the first accepted European balloon flight was made by the Montgolfier brothers in France.

As Spohrer and Woodman searched De Gusmao's records, they found his specifications for the ancient Peruvian balloons. They learned that the balloonists of the Nazca desert had stitched together large, four-sided triangular balloons and held them over a fire pit. As the fabric was inflated with the warm air from the intense wood flames, the balloon would begin to drift upward.

The two explorers found textile fragments in Peruvian tombs which they speculated might have been used to fashion the ancient balloons. Ravsen Industries in Sioux Falls, South Dakota, told Spohrer and Woodman that the textile samples were "superior to the material that is used today in balloon or parachute silk." Ravsen Industries copied the fabric and built *Condor I* for Spohrer and Woodman.

The December 15, 1975 issue of *Time* magazine reported on the International Explorers Society's flight of the *Condor*:

> . . . an odd contraption . . . with an 88-ft.-high envelope made from fabric that closely resembles materials recovered from Nazca gravesites. The balloon's lines and fastenings were made from native fibers, the boat-shaped gondola was woven from totora reeds picked by Indians from Peru's 2.4-mile-high Lake Titicaca.
>
> . . . Once released . . . Condor climbed quickly, reaching an altitude of 600 ft. in 30 seconds. Then, buffeted by brisk winds, it fell back to earth and hit with a thud that bounced the two pilots out of their gondola. Free of both pilots and ballast, Condor lifted off again, rose to 1,200 ft., flew about 2½ miles in 18 minutes, and then landed gently on the plain.
>
> . . . Michael DeBakey [an International Explorers Society

director] feels that the point has been made. "We set out to prove that the Nazcas had the skill, the materials and the need for flight. . . . I think we have succeeded."

For most modern men and women it does not really require a great leap of faith in the native intelligence of ancient man to suppose that he might have fashioned kites, balloons, and even rudimentary manned-gliders, but could it also be possible that he was able to place himself in the cockpit of heavier-than-air, motor-driven aircraft?

The late Ivan T. Sanderson became very excited by "little gold airplanes a thousand years old" which turned up in the collection of the Colombian National Collection, and later in half a dozen places, including the Chicago Natural History Museum and the Smithsonian Institution. Sanderson felt the South American artifacts were important because they represented neither animals, insects, birds, nor an ancient artisan's fantasies, but inanimate objects that "we could call out of this world."

Sanderson wrote an article about the airplane-shaped artifacts for a large-circulation, national men's magazine, but he was very disappointed in their choice of photographs of the models, which he felt weakened his case. In his own *Pursuit* newsletter, April, 1970, Sanderson firmly stated his position:

> The "discovery" of these little artifacts is probably one of the most pertinent ever made through archaeological enterprise . . . and it has become much more pertinent since more of these items have come to light and the aerodynamics engineers and designers have had a chance to analyze them. The original notion that they were "zoomorphic fantasies" . . . has now been completely demolished by the zoologists who, with all the will in the world . . . simply cannot come up with any animal that has the features of these items, while so many of these features are exactly and precisely that of airplanes. At least a possibility has to be faced . . . that somebody had airplanes *circa* 500 to 800 A.D. in northwestern South

America, and that local artists made models of them to the best of their ability and visibility.

The question as to who made the things that formed the models for these little pendants, presents quite another problem. There are three alternatives. Either there was a highly developed human civilization thereabouts at that time (or earlier); these things came out of the sea and were devices built by some underwater civilization; or they came down out of the skies from space and were subsidiary craft employed by intelligent entities from elsewhere visiting, surveying, or colonizing this planet. . . .

Not to be outdone by the ancient South Americans, Dr. Khalil Messiha believes that he has found evidence to indicate that the Egyptians had flying machines as early as the third or fourth century B.C. What is more, Dr. Messiha's brother, a flight engineer, agrees with him and adds that the aerofoil shape of the models discovered among some ancient bird figures demonstrates a "drag effect" evolved only recently after many years of aeronautical engineering research.

Dr. Messiha found the model glider or airplane in 1969 when he was looking through a box of bird models in one of the Cairo Museum's storerooms. The glider, made of sycamore wood, bears a striking resemblance, Dr. Messiha has since learned, to the American Hercules transport plane, which has a distinctive wing shape.

Most of the bird figures that have been found at excavations in Egypt are half-human, half-bird in design, but this object was very different. It seemed to be a model of a high-winged monoplane with a heart-shaped fuselage, which assumes a compressed elipse toward the tail.

"It is the tail that is really the most interesting thing which distinguishes this model from all others that have been discovered," Dr. Messiha was quoted in the May 18, 1972, issue of the *London Times*. The tail, it seems, has a vertical fin. There is no known bird that flies equipped with a rudder.

"No bird can produce such a contortion at the rear of its

body to assume anything that looks like the model. Further-more there is a groove under the fin for a tailplane [cross-piece] which is missing,'' the *Times* added.

In addition, as Dr. Messiha had learned from his several-year study of Egyptian bird figures, all other models had been lavishly decorated and had been fitted for legs. The glider has no legs and only very slight traces of an eye that had been painted on one side of the "nose," together with two faint reddish lines under the wing.

Dr. Messiha pointed out for Michael Frenchman of the *Times* that the ancient Egyptian engineers always made mod-els of contemporary things, from their funeral boats to their war chariots.

We know that funeral boats and chariots existed, because their full-scale versions have been found in addition to their models. Dr. Messiha has come to believe that the glider that he discovered in a box of relics excavated at Saqqara in 1898 is a scale model of a full-sized flying machine of some kind.

According to the *Times*, Dr. Messiha studied fine art for five years before he took up medicine. The 48-year-old doctor is also an illustrator and an engraver, and some years ago he received a prize for constructing model aircraft. "This glider is very much like some of the scale model planes I used to make 20 years ago,'' he said.

"This is no toy model,'' Dr. Messiha emphasized. "It is too scientifically designed and it took a lot of skill to make it.''

The doctor is presently engaged in research in ancient Egyptian sciences and engineering. Dr. Messiha believes that the Egyptians were very advanced in certain areas of tech-nology, "including elementary aeronautics.''

INFO Journal, Spring, 1973, found it "significant" that the glider had come to the Cairo Museum from the dig at Saqqara:

At Saqqara about 2700 B.C. the first Egyptian pyramid was built for the second king of the Third Dynasty, Neterkhet (or

Zoser, as the Greeks would later call him). Zoser's architect
. . . was credited by the later Egyptians as having invented
the art of building in hewn stone, as being a great astronomer,
magician, and the father of medicine. He came to be deified as
the son of the god Ptah, and the Greeks long afterward identified him with Asklepios, their own god of medicine.
   Saqqara was a shrine for millenia . . . . The little airplane
could have been the brainchild of Imhotep, dreaming of flight,
or of some other practical dreamer such as Archytas of Tarentum. Archytas (c. 400 b.c.) . . . is also alleged to have devised a flying machine in the form of a dove, propelled by compressed air. . . .

A. Neuburger's *Technical Arts and Sciences of the Ancients* agrees: "Archytas of Tarentum, about 400-365 b.c., set
in motion a flying machine in the form of a wooden dove by
means of compressed air." Archytas was a Greek philosopher
and a friend of Plato.

While Dr. Messiha credited the ancient Egyptians with
"elementary" aeronautical knowledge, G. R. Josyer, director
of the International Academy of Sanskrit Research in
Mysore, India, stated on September 25, 1952, that Indian
manuscripts several thousands of years old dealt with the construction of various types of aircraft for civil aviation and for
warfare.

The specific manuscript on aeronautics included plans for
three types of *vimanas* (aircraft), the *Rukma*, *Sundara*, and
Shakuna. Five hundred stanzas of an ancient text treat of
such intricate details as the choice and preparation of metals
which would be suitable for various parts of *vimanas* of different types.

Although the ancient manuscripts had been compiled by
venerable scribes and priests, the stanzas did not discuss the
mysticism of the Hindu philosophy of Atman or Brahman, but
detailed more mundane matters that the learned men of old
had considered essential for the "existence of man and the
progress of nations both in times of peace and war." These vi-

tal bits of information, according to Mr. Josyer, included the design of a helicopter-type cargo plane, specially constructed to carry combustibles and ammunition, and the drawings for double- and triple-decked passenger planes, capable of transporting as many as 500 persons.

There were eight chapters in the aeronautics manuscript that provided plans for the construction of aircraft that flew in the air, traveled under water, or floated pontoon-like on the water's surface. Some stanzas told of the qualifications and training of pilots. The ancient *vimanas* were equipped with cameras, radio, and a kind of radar system.

When word of the Sanskrit manuscripts reached beyond the monasteries in which they had been kept, the academy in Mysore began to receive a great deal of worldwide pressure to translate the ancient texts. Thus, at the age of eighty-one, Mr. Josyer "had to sit up and translate the technical Sanskrit into readable English, and scrutinize the printing of both the Sanskrit and the English."

The aged scholar found that the *Vymanika Shastra* consisted of "nearly 6,000 lines, or 3,000 verses of lucid Sanskrit dealing with the constructions of airplanes. That the vocabulary of ancient Sanskrit could in simple flowing verse depict the technical details with effortless ease is a tribute to the language—and the greatness of the author." (The work is attributed to Maharshi Bharadwaja, a Hindu sage who recorded the spiritual, intellectual, and scientific fields of ancient Indian civilization.)

The translation was finally published in book form by the Coronation Press of Mysore in 1973. In his foreword to the work, Mr. Josyer presents his assessment of the implications of the *Vymankia Shastra*:

> The 20th Century may be said to be made historic by two achievements: The bringing of Moon-rock from outer space and the publication of *Vymankia Shastra* from the unknown

past. . . . The *Vymankia Shastra* is a cornucopia of precious formulas for the manufacture of aeroplanes, which should make Lindbergh, Rolls, Zeppelin, De Havilland, Tupolev, and Harold Gray of Pan American, gape in astonishment, and if duly worked up, herald a new era of Aeroplane manufacture of the benefit of Mankind.

Ancient scientists insisted that any who would seek to pilot a *vimana* should acquaint himself thoroughly with 32 secrets of the working of the craft. As one examines the "secrets," he finds a peculiar mixture of technology and mysticism, almost as if pilot and craft achieve a kind of harmony and unity as one single, living entity. One should, for example, learn such techniques as the following in order to pilot a *vimana*:

Maantrika: The invoking of mantras which will permit one to achieve certain spiritual and hypnotic powers so that he can construct aeroplanes which cannot be destroyed.

Taantrika: By acquiring some of the tantric powers, one may endow his aircraft with those same powers.

Goodha: This secret permits the pilot to make his *vimana* invisible to his enemies. *Adrishya* accomplishes the same purpose by attracting "the force of the ethereal flow in the sky."

Paroksha: This helpful hint enables the pilot to paralyze other *vimanas* and put them out of action.

Aparoksha: One may employ this ability to project a beam of light in front of his craft to light his way.

Viroopa Karana: With this skill mastered, the pilot can produce "the thirty-second kind of smoke," charge it with "the light of the heat waves in the sky" and transform his craft into a "very fierce and terrifying shape" guaranteed to cause "utter fright to onlookers." *Roopaanara* can cause the *vimana* to assume such shapes as those of the lion, tiger, rhinoceros, serpent—even a mountain—to confuse observers.

Suroopa: If one can attract the thirteen kinds of "Karaka force," one can make the *vimana* appear to be "a heavenly damsel bedecked with flowers and jewels."

Pralaya: This deadly secret pushes electrical force through the "five-limbed aerial tube" so that the pilot may "destroy everything as in a cataclysm." *Vimukna* sends a poison powder through the air to produce "wholesale insensibility and coma."

Taara: This ability, once mastered, provides the pilot with another means of avoiding contact with an enemy or hiding his purpose from observers: "By mixing with ethereal force 10 parts of air force, 7 parts of water force, and 16 parts of solar glow, and projecting it by means of the star-faced mirror through the frontal tube of the *vimana*, the appearance of a star-spangled sky is created."

Saarpa-Gamana: This secret enables the pilot to attract the forces of air, join them with solar rays, and pass the mixture through the center of the craft so the *vimana* will "have a zigzagging motion like a serpent."

Roopaakarshana permits the pilot to see inside an enemy's airplane, while *Kriyaagrahana* allows one to spy on "all the activities going on down below on the ground."

Jalada roopa instructs the pilot in the correct proportions of certain chemicals which will envelop the *vimana* and give it the appearance of a cloud.

As the reader will see from even a casual reading of the above "secrets," piloting a *vimana* sounds in some instances very much like having an out-of-body experience or very much like applying psychokinesis, mind-over-matter, to an inanimate object and making it fly.

Did the *vimana* really fly by the magic of the mind to levitate it?

Were the *vimana* the legendary "flying carpets" of mysterious India?

Why was it necessary to cultivate "spiritual," "hypnotic," and "tantric" powers in order to pilot a *vimana*?

Students of the UFO enigma will also find many parallels

between the "secrets" of effective *vimana* piloting and the "behavior" of UFOs.

UFOs have often been reported to have become suddenly invisible.

Certain men and women claim to have experienced a temporary paralysis during a close encounter with UFOs.

Beams of light projected in front of a UFO have been observed by thousands of UFO percipients.

The zigzagging, serpentine motion has often been a facet of UFO flight.

And certain witnesses to UFO activity have sworn that the object has changed its shape before their eyes.

The "secret" that enables the *vimana* to appear to be a "heavenly damsel bedecked with flowers and jewels" might cause the student of UFO phenomena to wonder about a parallel between this *vimana* technique and the appearance of the Blessed Mother at Lourdes, Garabandal, and other sacred shrines coincident with the manifestation of "another sun" or a "glowing ball" in the sky. Likewise, the transformation of UFO to cloud has been reported by many observers.

Were the *vimana* psychological constructs employed to enable one to have certain types of visionary, out-of-body experiences, the physical ancestors of the UFOs we are still observing in our own skies, or were they truly material craft carefully built and judiciously flown by ancient aviators in India?

The venerable Maharshi Bharadwaja, author of the *Vimankia Shastra*, begins his work by making obeisance to the Divine Being, "who is visible on the crest of the Vedas, who is the fountain of eternal bliss, and whose abode is reached by *vimanas* or areoplanes." One does not think of flying to heaven in a physical craft. Again, the *vimana* seems almost to be another term for "astral" or "soul" body.

"Having studied the *Shastras* or sciences propounded by

previous men of science to the best of my ability for the benefit of mankind," the author continues, "I shall deal with the science of aeronautics, which is the essence of the Vedas, which will be a source of joy and benefit to humanity, which will facilitate comfortable travel in the sky from world to world. . . ."

If mastering the flight of the *vimana* will permit one to travel between worlds or dimensions of reality, then its function seems to be far beyond what one would normally expect of an "aeroplane," whether material craft or spiritual exercise.

Mr. Josyer may be correct when he declares the discovery of the *Vimankia Shastra* to be one of the historic achievements of the twentieth century. And we can but pause to wonder how many other vital secrets of technological and spiritual advancement may have been lost by forgotten civilizations and cultures.

CHAPTER SIX

Lost Civilizations and Vanished Peoples

In the minds of the orthodox archaeologist and anthropologist, the origins of the Amerindian have been clearly charted: Sporadic migrations of people crossed from Siberia into Alaska over a period of 15,000 years or more. Before these Siberians invaded the New World, the continent was like the Garden of Eden without an Adam or an Eve.

Most of the Amerindians with whom I have discussed the matter of the origins of the Indian people upon this continent simply do not accept the Bering Strait invasion theory as the true account of their genesis. They do not deny that such a Mongolian invasion took place any more than they deny that the European invasion took place, but they do not believe that their origins were in Siberia any more than they were in England or France.

The physical makeup of certain tribes, especially those of the far West, bear mute testimony that interbreeding occurred between Amerindian and Mongolian types, just as nearly all tribes bear evidence of interbreeding with European types.

But the traditional Amerindian stoutly maintains that just as they were here to welcome the European, so were various tribes already on these shores to greet the Siberian.

The Amerindian peoples, the traditionalist believes, are the descendents of those who survived the destruction of a great civilization that once existed on this continent. And he believed this long before the sleeping prophet Edgar Cayce declared that the inhabitants of lost Atlantis were a red-skinned people.

For one thing, the people who came over from Siberia were primitive. How then, utilizing this orthodox timetable, does one explain the sophisticated stone tools, found in an ancient Mexican stream bed, which were considerably more advanced than those used in Europe and Asia 250,000 years ago? The most primitive tools found in the stream bed were of a type used in the Old World 35,000 to 40,000 years ago.

"We have apparently found geological data that led to a head-on confrontation with apparently sound archaeological data," Dr. Ronald Frywell of Washington State University was quoted in *The New York Times*, November 18, 1973.

People have been discovering data that simply will not fit the accepted theories for the origin of the Amerindian no matter how much one bends, folds, or mutilates. How long these theories will be allowed to remain sacrosanct after such "head-on confrontations" is a question that begs a forthright answer.

For example, even more dramatic than finding sophisticated stone tools that may be 250,000 years old, are the various walled cities and fortifications that have been found scattered throughout the United States.

In Rockwall, Texas' smallest county, four square miles support the great stone walls of an ancient fortification—some of which reach heights of forty-nine feet. The walls are about eight inches thick. The stones have been placed on top of each other with the ends breaking near the center of the stone above or below, just as a fine mason would build a wall. The stones give the appearance of having been beveled around their edges.

Raymond B. Cameron told Frank Tolbert, columnist for the *Dallas Morning News*, that "four large stones taken from wall segments appear to have been inscribed by some form of writing."

In the 1920s, a visiting archaeologist declared that the walled city appeared remarkably similar to the buried cities that he had excavated in North Africa and the Middle East.

Cherokee Indians, ancient Welshmen, and a vanished tribe of moon-eyed, blond whites have been at various times credited with the construction of the 885-foot wall on Fort Mountain in northern Georgia. The wall runs from east to west and, at various intervals, looms over 29 pits that look as though they might have served as ancient foxholes from which defenders repelled invaders. The wall ranges from a height of seven feet to only two or three feet. The quantity of rock along the line of the wall suggests that at one time the wall might have been much higher.

According to John Fleming, writing in *Southern Living*, December, 1969: "The lack of warlike artifacts (or any artifacts for that matter) lends credence to the old Cherokee legend that the wall was built by a race of white people who worshipped the sun. This, according to the tale, is the reason the wall was built from east to west—from the rising sun to the setting sun.

"But if this is true, then what happened to the sun worshippers? How did they leave the area without leaving a trace of artifacts or funerary remains?"

On June 27, 1969, workmen leveling a rock shelf at 122nd Street on the Broadway Extension between Edmond and Oklahoma City, Oklahoma, uncovered a rock formation that created a great deal of controversy among investigating authorities. To a layman, the site looked like an inlaid mosaic tile floor.

"I am sure this was man-made because the stones are placed in perfect sets of parallel lines which intersect to form a diamond shape, all pointing to the east," said Durwood Pate, an Oklahoma City geologist who studied the site and was quoted in the *Edmond Booster*, July 3, 1969. "We found post holes which measure a perfect two rods from the other two. The top of the stone is very smooth, and if you lift one of them, you will find it is very jagged, which indicates wear on the surface. Everything is too well placed to be a natural formation."

Delbert Smith, a geologist, president of the Oklahoma Seismograph Company, said the formation, which was discovered about three feet beneath the surface, appeared to cover several thousand square feet. The *Tulsa World*, June 29, 1969, quoted Smith as saying: "There is no question about it. It has been laid there, but I have no idea by whom."

In 1973 and 1974, the following confrontations with the orthodox timetable for the beginnings of human habitation on this continent were recorded by archaeologists:

"The amino-acid dating of a human skull found in California indicates human habitation of North America 50,000 years ago. . . ."

"The oldest discovered site of human habitation east of the Mississippi—14,000 to 15,000 years—was uncovered south of Pittsburgh."

"The mysterious Medicine Wheel in Wyoming's Bighorn Mountains was discovered to have astronomical alignments and may have been used as an observatory by nomadic Plains Indians. . . ."

"A carved bone hide scraper, found on the Old Crew River, a few miles from the Alaskan border, was radiocarbon-dated at between 25,000 and 32,000 years. . . ."

Stanley Taylor of the Films for Christ Association superintended the excavation of the Paluxy River bank to uncover fresh tracks of man *in situ* with dinosaur tracks.

Three-toed dinosaur tracks along the Paluxy River near Glen Rose, Texas.

Dr. C.L. Burdick.

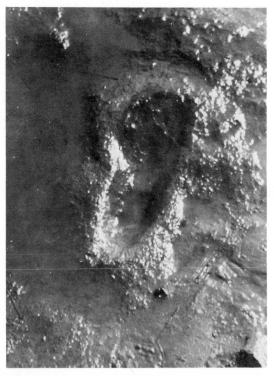

One of the man tracks that Stanley Taylor uncovered along the Paluxy River bank. Note the heel and instep. "Whoever" made this track apparently did so when the mud was soft. Taylor exposed several such tracks, both dinosaur and, apparently, human.

Dr. C.L. Burdick.

Dr. C.L. Burdick.

These man tracks were found *in situ* with the tracks of Tyrannosaurus Rex, the most vicious land animal the world has ever known. The man tracks are about seventeen inches in length. Roland Bird of the American Museum of Natural History admitted the human tracks were "perfect in every detail." But when he found the tracks associated with dinosaur prints, he disputed his first identification.

The tracks were jackhammered out of the Paluxy River a few miles northwest of Glen Rose, Texas, and taken to the Berry roadside museum on Highway 66. Here, a Dr. Westcott of Michigan pronounced them genuine, calling attention to the high instep which suggests a walking motion.

Dr. Clifford Burdick has spent more than thirty years in a study of what appear to be human footprints in strata contemporaneous with dinosaur tracks. Early in 1975, Dr. Stanley Rhine of the University of New Mexico announced his discovery of humanlike footprints in strata indicative of being 40 million years old.

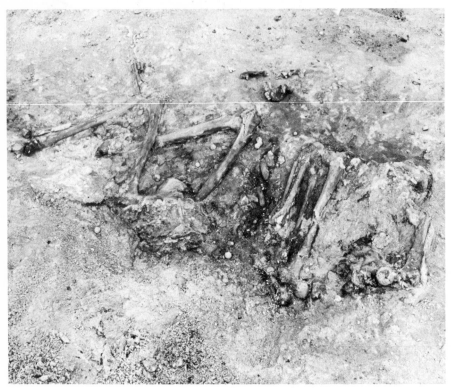

Department of Anthropology, University of Utah.

In May of 1971, tour guide and amateur geologist-archaeologist Lin Ottinger found traces of human remains in a geological stratum approximately 100 million years old. The find was made at the Big Indian Copper Mine in Lisbon Valley, about 35 miles south of Moab, Utah.

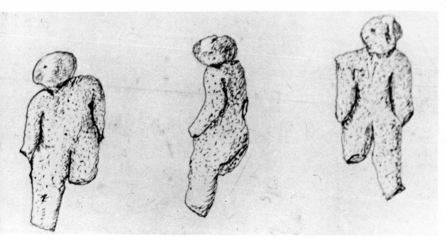

AMERICAN GEOLOGIST, 1889.

In August, 1889, near Nampa, Idaho, M.A. Kurtz picked up an odd-looking lump of clay that had been brought up from a depth of 300 feet during a well-drilling operation. When he broke it open, he discovered what he thought looked like a tiny human figure made of clay. The controversy over the apparent antiquity of the Nampa Image has raged ever since. If it were possible to apply the radioactive carbon technique for determining age to nonorganic materials, the mystery might be solved. But if the object were found to be manmade, its origins would present an even deeper, more controversial enigma.

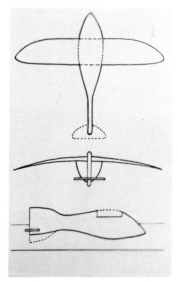

PURSUIT Newsletter.

Dr. Khalil Messiha believes that he has found evidence to indicate that the Egyptians had flying machines as early as the third or fourth century B.C. Dr. Messiha found a model glider or airplane in 1969 when he was looking through a box of bird models in one of Cairo Museum's storerooms. He is convinced—and his brother, a flight engineer, agrees with him—that the glider, made of sycamore wood, bears a striking resemblance to the American Hercules transport plane.

MILTON R. SWANSON

"The Museum of Fine Arts in Boston has the world's finest and most complete laboratory, which was built in cooperation with M.I.T. I was able to have them run it through every kind of test for two years. Still no answer as to its period or origin."

In its June, 1851, issue the *Scientific American* carried an item about a metallic vessel that had been blasted out of an "immense mass of rock" when workmen were excavating on Meeting House Hill in Dorchester, Massachusetts: "On putting the two parts together, it formed a bell-shaped vessel, 4½ inches high, 6½ inches at the base, 2½ inches at the top, and about an eighth of an inch in thickness. The body of this vessel resembles zinc in color, or a composition metal, in which there is a considerable portion of silver. On the sides there are six figures of a flower, or bouquet, beautifully inlaid with pure silver, and around the lower part of the vessel a vine, or wreath, inlaid also with silver. The chasing, carving, and inlaying are exquisitely done by the art of some cunning workman. This curious and unknown vessel was blown out of the solid pudding stone, fifteen feet below the surface. . . . Dr. J.V.C. Smith, who has recently travelled in the East, and examined hundreds of curious domestic utensils . . . has never seen anything resembling this. . . . There is no doubt but that this curiosity was blown out of the rock. . . ."

Recently, the present owner of this curious artifact wrote to Brad Steiger and informed him that the vessel is still unidentified after over one hundred years. According to Milton Swanson of Maine: "It had been given to Harvard College, but because of its mysterious origin they relegated it to a closet. The building supervisor finally brought it home to Medford, Mass. He sold it to me just before he died in his eighties.

"Through the years I have had so-called experts look at it, and no one ever came up with an answer. Its age and use is just unexplainable. It is almost black, but the metal is composed of brass with zinc, iron, and lead. The inlay is pure silver, and I had to put lacquer on to protect it. I always felt that it was a burial ash container.

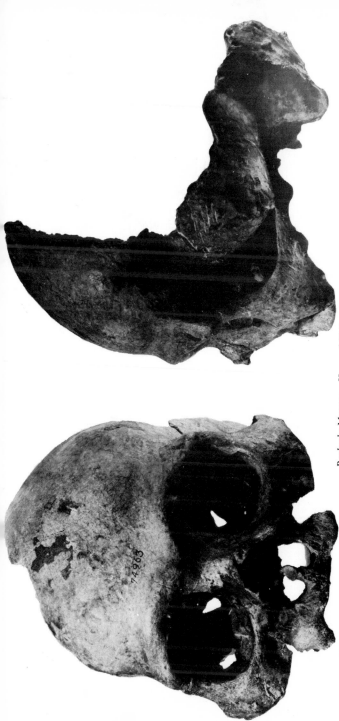

On February 15, 1866, at a depth of 130 feet, James Mattison's gold mine shaft struck what he believed to be a root of a petrified tree. Once the lime deposits had been cleaned from the object, J.C. Scribner, a merchant of Angel's Camp, California, was startled to find that he held a human skull in his hands. If the skull is that of a man who lived in the Pliocene period, then it constitutes not only the oldest fossilized human remains on the North American continent, but in the entire world as well. Few scientists deny that the skull has great antiquity, but they point out that only a few hundred years in the deep, limestone caves of Calaveras could produce the chemical changes of fossilization.

Greek National Archaeological Museum, Athens.

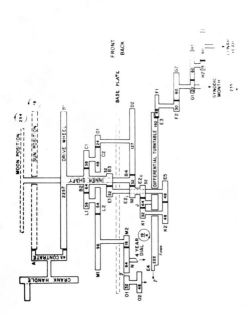

Shortly before Easter, 1900, a group of sponge fishers brought up a peculiar mechanism from the wreck of an ancient Greek ship off the islet of Antikythera. It was not until half a century later that Dr. Derek DeSolla Price, Avalon Professor of History at Yale University, made the astonishing discovery that the mechanism was "a calendric Sun and Moon computing mechanism which may have been made about 87 B.C." DeSolla Price states that we have drastically underestimated the "whole story of Greek science."

The Antikythera device is ". . . an elegant demonstration or simulation of the heavens . . . uses fixed gear ratios to make these calculations of the soli-lunar calendar, and it does this more by using pointer readings on a digital dial than by causing a direct geometrical modeling of the paths of the planets in space."

Certain archaeologists and anthropologists might have to swallow a pet theory or two, but most would be able to accept the above discoveries with a modicum of pain. After all, 50,000 years can be endured, but the suggestion of sentient humans walking about writing on North American walls during the Carboniferous Era, 250 million years ago, simply subjects the orthodox thinking apparatus to more shocks than may be comfortably sustained.

But consider the following footprints in the geologic strata of time:

In the early 1930s Dr. Wilbur Greely Burroughs, head of the Geology Department of Berea College, was guided to a site in the Kentucky hills where he was able to locate ten complete manlike tracks and parts of several more in Carboniferous sandstone. All the accumulated evidence indicates that they were impressed upon a sandy beach in the Pennsylvanian Period of the Paleozoic Era—which dates the humanoid impressions somewhere around *250 million* years ago. Dr. Burroughs kept his work secret for seven years. One can imagine that he wanted every opportunity to study the tracks of a bipedal creature that could have worn size 7½ EE shoes.

"Three pairs of tracks show both left and right footprints," Dr. Burroughs told Kent Previette of the Louisville *Courier-Journal* magazine some years later (May 24, 1953). "Of these, two pairs show the left foot advanced relative to the right. The position of the feet is the same as that of a person. The distance from heel to heel is 18 inches. One pair shows the feet about parallel to each other, the distance between the feet being the same as that of a normal human being."

The Pennsylvania Period was the age of giant amphibians. Could the tracks have been caused by one of them?

Dr. Burroughs thought it unlikely. "There is no indication of front feet, though the rock is large enough to have shown front feet if they had been used in walking." Dr. Burroughs was emphatic that the creatures, whatever they might have been, walked on their hind legs. Nowhere on the site were there signs of belly or tail marks.

On May 25, 1969, the *Tulsa Sunday World* carried an article describing fossilized footprints found by Troy Johnson, a North American Rockwell liaison engineer. Just a few miles beyond Tulsa's eastern city limits, Johnson removed earth, roots, and stone from an outcropping of sandstone to reveal animal prints—many of which he could not identify—and some distinctive, five-toed, humanlike footprints.

C. H. McKennon of the *Tulsa Sunday World* presented Troy Johnson's quiet arguments in favor of the footprints' authenticity: "The chunk of sandstone containing the big prints is a massive weight of an estimated 15 tons, which rules out the possibility of someone transporting it to the top of the hill. Also, the stone is of the same strata as other specimens of sandstone dotting the hilltop, indicating there was a monumental 'uplift' of the earth's crust ages ago. . . ."

For those who are skeptical that a prehistoric civilization could have thrived on our own continent and left but the scantest vestiges of its culture to alert future generations of its existence, let us consider what would happen if a catastrophe should decimate our own civilization. What would remain for archaeologists to unearth 15,000 years from now?

We are builders in wood and metal. Our most majestic stone buildings are little more than façades supported by thin tendons of steel. In a thousand years, even without flood, fire or nuclear warfare, our major cities would be little more than rubble. Our complex superhighways would be crumbled bits of hardness beneath layers of vegetation. Our once intricate railway system would be red dust blowing in the wind.

If volcanic lava and dust should happen to blanket a major city in a sudden eruption—such as the violent bursting of Vesuvius that preserved Herculaneum and Pompeii—a portion of our civilization would be preserved as if in a gigantic museum display.

But if we were to enter another ice age and enormous glaciers should creep down from the north, as they have several times previously in the past million years, everything in their inexorable path would be pulverized. One such glacier would be enough to wipe out any trace of our civilization. Perhaps only scattered pieces of porcelain would persist to inspire future scholars to write doctoral dissertations on what manner of priesthood served a deity at the altar of the flush toilet.

In my own thinking, I have brought the matter down to the following, personal analogy: Although the Iowa community in which I formerly resided is small and several hours away from any large, metropolitan area, it possesses all the modern conveniences, along with up-to-date shops and supermarkets, a well-staffed hospital, and a small college. Let us hypothesize the unpleasant situation of the entire civilized world blasting itself with nuclear bombs in an insane orgy of destruction. All the major cities would be blitzed to nothingness, but life in small communities, such as the one I am describing, would do its best to continue.

The television set would no longer be functional, except, perhaps, as something on which to stack books.

The radio would be functional only until the local station suffered a breakdown, which would necessitate ordering new parts. It would be impossible, of course, to order new parts for anything from anywhere.

Automobiles would be functional only until the storage tanks of gasoline had been exhausted, and the local mechanics could no longer provide repairs.

The doctors at the hospital and the clinics would do their best to instruct the more intelligent members of the community in the rudiments of modern medicine; but modern medicine's magic is very weak without its attendant technology, which would now have been destroyed.

The teachers and professors at the public schools and the colleges would surely do their best to keep alive the ideals of our culture; but effective crop raising would now seem much more important than philosophy. Survival would take priority over Shakespeare.

One day the last machine would break down, and there would be no one who remembered how to repair it. The X-ray machines, the radios, the long-dead dry-cell batteries, although still revered, would be useless. They would soon be forgotten as actual implements, but they would be elevated to the status of magical artifacts in the legends of the Iowa villagers, as survival instinct would direct them back into the rapidly encroaching forests.

Someday, perhaps 20,000 years in the future, someone might "remember" how to use the marvelous machines—the box that could see other people thousands of miles away; the room that could move inside big houses; the chariot that could fly through the clouds. Or perhaps one day more technologically sophisticated men and women from across the Big Water would declare the descendants of those Iowa villagers to be primitive, aboriginal people of the New World.

Since the concept of "ancient astronauts" has become a popular one in the last two years, it would be only fair to consider briefly that the various artifacts which we have discussed in this chapter might have been "seeded" on this continent by colonists from other worlds or other dimensions.

It is also conceivable that the collective unconscious of the Amerindians may recall their origin in a culture which might be found in some other planet, rather than a "lost civilization" on Earth. Or that same collective unconscious may harbor memories of the intimate interaction of their ancestors with visitors from extraterrestrial ports of call.

Just as Great Britain had its Stonehenge, Egypt its pyramids, and the Mayans their temples, all of which served as giant calendars as well as impressive monuments, the nomadic Plains Indians of North America had their Big Horn Medicine Wheel to signal the summer solstice—or, perhaps, the "gods."

Just above the timberline in the Big Horn Mountains of northern Wyoming, the Medicine Wheel's pattern of stones etches an imperfect circle with a diameter of about 25 meters. A cairn of stones about four meters in diameter establishes the hub of the wheel. Twenty-eight "spokes" issue from the hub and connect with the outer rim.

The Big Horn mountains held special significance for the Crow, the Sioux, the Arapaho, the Shoshone, or the Cheyenne—any of whom might have erected the wheel—but none of these tribes were known for building any kind of stone monuments. Bits of wood found in one of the six smaller cairns situated unevenly about the rim indicates that the Medicine Wheel has been there since at least 1760. The monument has been known to white men for well over a hundred years, but conjecture about its true purposes has only inspired mysteries and tall tales.

In the June 7, 1974, issue of *Science*, astronomer John A. Eddy of the High Altitude Observatory in Boulder, Colorado, states that two summers' research have convinced him that the Big Horn Monument may well have been a primitive astronomical observatory that served its creators at least as well as Stonehenge served its primitive astronomers. The high altitude (9,640 feet) and the clear horizons of the monument make easily visible the marking of sunrise and sunset at the summer solstice. The accurate knowledge of the first day of summer would have been a most important bit of intelligence for a nomadic people whose very lives depended on astute awareness of seasonal changes.

There are numerous Amerindian legends that suggest an interaction between native American peoples and Star Dwellers. Nearly every tribe has its accounts of "Sky Ropes"— ropes of feathers that permitted People From Above to come to the Earth Mother and, on occasion, enabled men and women to fly upward. Along with the stories of magical ropes are tales of flying canoes, airships, and moons that descend to Earth.

Many Amerindian tribes believed that the stars were the homes of higher beings who had a connection with, and a mysterious relationship to, humans. Others held that the stars were themselves actual ministering intelligences.

Numerous tribes had accounts of warriors who had found themselves enamored of Star Wives and of tribeswomen who had been enticed by Star Husbands. Often the Amerindians found "magic circles" that the Star People had burned into the grass, just as their European brothers across the Big Water were finding "fairy circles" that the dancing wee ones had tromped into the meadows during their nocturnal revels.

The Chippewa have a legend that tells of a great "star with wings" that hovered over the tops of the trees. Some of the wise men thought of it as a precursor of good; others, understandably, feared the star and saw it as the forerunner of terrible times.

The star hovered near the village for nearly one moon (month) when a Star Maiden approached a young warrior and told him that she was of the winged star. They had returned from a faraway place to this, the land of their forefathers, and they loved the happy race they saw living in the villages. The star, she said, wished to live among them.

The warrior told the council of his visitation, and representatives went to welcome the Star People with sweet-scented herbs in their pipes of peace. The winged star stayed with them for only a brief time, however, before it left again to live in the southern sky. As a token of its eternal love, according to the Chippewa, the Star People left the white water lily on the surface of the lakes.

Each of the Amerindian tribes with which I am familiar cherishes legends that tell of their people rising from the destruction that had been visited upon a former civilization. The majority of the accounts deal with the surviving peoples having escaped from a terrible flood, which immediately suggests both the biblical story of The Deluge and the Atlantis mythos.

The principal point of each of the Amerindian myths of destruction and rebirth is that civilization is cyclical, continually being born, struggling toward a golden age, then slipping backward in moral morass, forward into its death throes . . . only to be reborn so that the process may begin once more.

Immanuel Velikovsky once said in an interview with *Science and Mechanics* magazine that prior civilizations are buried so deeply within the lower strata of the Earth that we simply do not have archaeological evidence of their existence.

"But we have abundant references in literature—even in rabbinical literature—that many times . . . before this present Earth Age existed, the *same* Earth was created—then it was leveled and *re*created; all civilizations were buried," Velikovsky stated.

"By far the vast majority of ancient texts deal specifically with the phenomenon of catastrophism. . . . In the Old Testament we read of geological disturbances in which a mountain melts like wax, the sea being torn apart or erupting on the land, cosmic debris bombarding the people, and the ocean parting to show the foundations of Earth—and we say all these things are metaphors. This is what makes it appear to me that mankind is a victim of collective amnesia. As such a victim, he likes to play with atomic weapons, then repeat the events that took place! The victim of amnesia who has undergone a traumatic experience seems to want to relive those experiences."

The Seneca legend of the Seven Worlds says that man has relived such "traumatic experiences" six times before and that we stand on the brink of destruction prior to entering the final world in our evolutionary cycle.

The Hopi legend of the Four Worlds agrees and states that mankind is about to enter the final world after a last great war—a war that shall resolve the spiritual with the material and create one world under the power of the Creator.

For the Amerindian traditionalist, the destructions of the

103

previous worlds have been a necessary part of mankind's spiritual evolution. Because man has repeatedly forgotten the lessons of the Great Spirit, the Earth Mother has been periodically cleansed for new epochs. If the old prophecies are correct, we may have little time to avoid becoming part of the collective amnesia of some future generation.

If civilization has been cyclical, then it would seem perfectly possible for the Creationists to have their paradigm of a Special Creation and a God-directed Time that began in 4004 B.C. and for the Evolutionists to have their paradigm of a continuing creation that began billions of years ago. The Evolutionists would have to compromise and permit an occasional catastrophe, and the Creationists might have to concede at least two or three thousand more years to their starting date.

The most persistent, haunting memory of a world before our own is given periodic physical expression in the almost ritualistic searches for lost Atlantis. Traditionally, Atlantis was a continent in the Atlantic ocean that was shaken by a series of violent cataclysms, which brought it below the surface of the waters.

So desperately has mankind appeared to have suffered this loss that some years ago a public opinion poll found that the people in the United States would rank the discovery of Atlantis as a greater news story than the second coming of Christ. The majority of orthodox scientists, however, evaluate the legends of Atlantis as having been but philosophical parables told by Plato in order to provide a dramatic object lesson in civics for his students.

The great blueprint for Atlantean research of a somewhat less than orthodox scientific bent was laid by Ignatius Donnelly in his classic work *Atlantis*. Donnelly had been a Republican congressman from Minnesota who elected to spend his spare time while in Washingon, D.C., researching any literary or archaeological erratics that could in any way be associated with the myth of Atlantis.

In his book, Donnelly makes a strong case for a common source, Atlantis, having seeded the civilizations on both sides of the Atlantic. He details with profuse illustrative examples that cultures in Europe, Africa, Asia, Central and South America have almost precisely the same arts, sciences, religious beliefs, social customs, personal habits, and folk traditions—from pyramids to palaces, from metallurgy to monetary systems, from public works to techniques of warfare, from the belief in the immortality of the human soul to the belief in ghosts and such supernatural companions as fairies and "wee people."

Donnelly maintained that Atlantis was a large island whereon man first rose from a barbaric state to a civilization that would not be equaled until 6,000 years after its destruction in 8,000 B.C. All the arts and mental refinements necessary to compose the essential attitude of civilization date back to the time of Atlantis.

The empire of Atlantis would not be matched even by the mighty, sprawling kingdoms of Rome or Great Britain. The government of Atlantis touched the minds of men and exacted homage from the fruit of their lands from the Gulf of Mexico to the Mississippi River, from the Amazon River to the Pacific coast of South America; from the Mediterranean to the west coast of Europe and Africa, the Baltic, the Black Sea, and the Caspian Sea. Egypt, the first major Atlantean colony, reproduced the civilization of the ancient island kingdom.

Those larger-than-life men and women who were regarded as gods and goddesses by the ancient peoples of the Mediterranean were in reality the kings, queens, and heroes of Atlantis. The supernatural acts that were attributed to those god-kings were merely an exaggerated recounting by primitive Greeks and Cretans of actual deeds performed by Atlanteans.

Atlantis was destroyed due to some terrible natural catastrophe, and the island kingdom sank beneath the ocean. Virtually the entire population of the nation perished, with the exception of the merchant marine and those who managed to escape in boats and rafts. The survivors of the greatest disaster known to the human race spread the account of the death of Atlantis to all of the nation's colonies, and the story has been passed to our own time and among all peoples as the memory of the Great Deluge.

The entranced clairvoyant Edgar Cayce envisioned Atlantis enduring two great cataclysms before a final catastrophe destroyed the island kingdom and transformed it from life to legend. The First Destruction ripped Atlantis, then a continent, with violent seismographic disturbances about 50,000 B.C. The Second Destruction broke up the land mass of Atlantis into five major islands in approximately 28,000 B.C. The Final Destruction, which plagued the great Sea Kings beneath the ocean, took place around 10,000 B.C.

In July, 1973, a group composed of educators, students, scientists, parapsychologists, and psychics embarked on a six-week expedition in search of Atlantis off Cadiz, Spain. Leading the expedition was Maxine Asher (who at that time served as educational consultant to Pepperdine University in California and who is still head of the Ancient Mediterranean Research Association) and her codirector, Dr. Julian Nava, vice-chairman of the Los Angeles City Board of Education and a professor of history at California State University, Northridge.

When I interviewed the attractive, enthusiastic Ms. Asher shortly before her early departure for Spain, she informed me that Dr. Manson Valentine, Gail Cayce (granddaughter of the famous clairvoyant), and Edgerton Sykes, a noted Atlantean scholar, would be accompanying her party. In addition, she said, they had been promised assistance from the Scripps Institute of Oceanography in San Diego and cooperation from the Spanish Ministry of Education and Science. In that same spirit of cooperation, Ms. Asher patiently answered my questions while she attended to a multitude of last-minute details.

"Your expedition is seeking Atlantis off the coast of Spain. What do you believe the perimeters of the continent to have been?"

Maxine Asher: I believe they extended as far south as the Canary Islands, on the back side of the Atlantic, and as far north as Ireland. I can't say *where* in Ireland, but I suspect as far as Galway, County Cork, and maybe a little farther. I think that Atlantis spanned the entire Atlantic Ocean at one time, perhaps one million or two million or more years ago; it's hard to date it at this point.

I believe that its northernmost boundaries on the other side were perhaps Nova Scotia, and the southernmost boundaries, Bimini, and maybe farther down. I know that everyone is saying, Yucatan, Peru; but I'm really not sure that it went too much farther than, say, Venezuela. Surely, its people could have migrated, but I don't see its perimeters extending that far.

Atlantis probably went down originally as a result of seismic upheavals under the ocean, but I do believe the final destruction was probably cosmically ordained. Hypothesizing, of course, the final destruction of Atlantis could have come cosmically, because the people had become so evil and had generated enough negative force that they had disrupted the Cosmos. Divine Retribution could have been the springboard—let me call it that—for the cosmic upheaval that destroyed Atlantis.

"What level of civilization would you say the Atlanteans had attained?"

Ms. Asher: Well, I think that's like asking what were the Americans like. We need to look at the Colonial Period and the Civil War. I would say that, technologically, around 50,000 B.C., they were very, very advanced. They had air travel, and they had underwater devices. They were very "modern."

"Comparable to our contemporary civilization?"

Ms. Asher: Comparable, but I don't think they were beyond. I don't think you can reconstruct Atlantis and find another United States, however. I think it probably was a different type of culture.

It is difficult to evaluate just what "advanced" is. When the pioneers traveled across the early United States, they said the Indians were in a less advanced state of civilization—but by whose standards?

The Atlanteans, I believe, did have a highly advanced technology; but I believe their technology was psychically oriented. They combined the psychic and the rational in such a molding that they were able to do some very impossible things, such as moving huge rocks for pyramids through psychokinesis. They read each other telepathically. In its most advanced period, life was probably very much governed by a combination of the best of two worlds—and they lived in harmony. I think we need now to look at what happened to them because, according to many accounts, they faced a decline that probably came about as a result of forces that were negative.

"Do you believe they may have had nuclear power?"

Ms. Asher: I think they had a form of energy as powerful as our nuclear energy, but I believe that it was psychic in nature.

I don't have to tell you, Brad, about the force of the psychic. If enough people get together for bad, using mind control, they can do an awful lot of damage.

"Why are you led to Cadiz? Why not Nova Scotia?"

Ms. Asher: Well, first of all, when I was doing my Masters' work in history, I spent much time trying to find the origin of the Etruscans. [Ms. Asher has, since this interview, received her doctorate.] I went to the Etruscan tombs and so forth, and nothing made any sense.

So I went to Crete, and there I became bothered by the origin of Linear A writing.

I scooted up to the Pyrenees Mountains, and I got bugged to death about Cro-Magnon man.

I began to tear around, looking for common origins, not even thinking of Atlantis.

I was led, I am sure, by the Divine Hand to Spain.

When I got to Cadiz—Plato had spoken about Atlantis lying beyond the Gates of Hercules—the vibrations were so strong, Brad, I thought I was going to jump out of my skin.

The first person I met was a taxi driver, who went into an entire dissertation on Atlantis. One thing led to another. I interviewed many people, and I said, "By God, I think Plato knew what he was talking about!"

I did extensive research on Cadiz, and I went to the university and they accepted my program. Not so much on Atlantis, but on the search for it. All knowledge, they stated, is transient; all knowledge is changing; and the finding of Atlantis will only be a vehicle, a doorway, that will open up a whole, brave New World for many of us.

I think Cadiz is a logical place to begin, but I suspect that we will be uncovering pieces of Atlantis all the way from Cadiz to Ireland. I'll be doing research in the Erin Islands in Ireland shortly because the Irish connection may, in fact, be more important than the Spanish, although all of it will be meaningful.

"Do you feel any currently existing lands were once a part of Atlantis?"

Ms. Asher: I think part of the Iberian peninsula was, specifically the area near the Guadalquiver River. I think that area could have been Tartessos and that there could have been a land bridge of sorts connecting that point to the Atlantean island.

I think Edgar Cayce was mistaken—again, I am hypothesizing—when he said that Poseidia was near Bimini. He said that the Atlanteans buried some of their records in the Pyrenees

Mountains, in Egypt, and also in Yucatan; but I still suspect that Poseidia, bearing the name of the Greek god of the sea, was the last little island near Cadiz, and that the Atlanteans brought their records to the obvious place close at hand.

"Do you believe that Atlantis is the true name of the lost continent?"

Ms. Asher: I don't know, but I belive that it had the combination of letters *Ata* or *Atla* in it.

"Those combinations are found often in Central and South America. Do you think that we are the descendants of Atlantean survivors?"

Ms. Asher: Yes, but other than the one boat with Noah and the animals, or whatever. I think that there may have been twelve boats, corresponding to the twelve tribes of Israel.

"And couldn't there have been Atlantean colonies in other lands?"

Ms. Asher: Of course. And I think some people began to leave early. I believe that people began to see the beginning of the end, you know, and took off. But you also remember that in the biblical story, only a few people believed Noah when he said that there was any sort of danger. That's why we have such isolated pockets of Atlanteans. We have the Basques in Portugal, the Gaelic Irish, and others.

"Do you think that we could suffer such a cataclysm in our time?"

Ms. Asher: Yes, but it may not be physical. It may be that the world will face some sort of catastrophe that will be psychic in nature, and only those who have learned to tap all levels of consciousness will survive.

The expedition of the Ancient Mediterranean Research Assocation terminated their search for Atlantis in mid-July, 1973, under a dark cloud filled with confusing rumblings. Reports were issued that Dr. Julian Nava had resigned in disgust because the Cadiz-based team had "blown it" with a premature dive and an exaggerated appraisal of the worth of a find.

Dr. Nava was quoted as saying that on the night of July 16th, just a few days before the extremely hard-to-obtain Spanish Land Permit was to be awarded to AMRA, an authorized dive had been made. When divers released a claim of the "greatest discovery in world history," the disgruntled Spanish government had denied the permit. Later, Dr. Nava clarified his position that he had resigned as codirector of AMRA only because of the necessity of his returning to the United States in order to meet publishers' deadlines on some textbooks he was writing.

The expedition had begun to take on overtones of a James Bond thriller, complete with shadings of international intrigue. (Ms. Asher did, in fact, write a book about the "Atlantis Conspiracy.") Ms. Asher left Spain with a group of students for the comparative quiet of Ireland.

Then, in September, 1973, reports from Cadiz indicated that several Spanish archaeological groups were searching for the site of the Atlantis "find" claimed by the Ancient Mediterranean Research Association on July 16th. AMRA Director Maxine Asher stated that her group was barred from using its official underwater archaeological permits and could not release its photographs or maps of the site until politics in Spain had stabilized.

"It is unfortunate," stated Dr. Asher, "that the scientific and educational interests of the entire world in the matter of Atlantis are thwarted by international intrigue."

An interview in the newspaper *Diario de Cadiz* on September 2, 1973, quoted archaeologist Jesus Aguero as saying, "One thing we cannot doubt is that the city of Atlantis exists . . . we even have Atlantean money, which can be found in the Louvre in Paris."

Professor Aguero further validated that vestiges of Atlantis can be found in the coastal waters north of Gibraltar, the area where the AMRA group made its finds.

Perhaps those who search for the lost Atlantis would do

best to concentrate *inland* on the North American continent, rather than plumbing the depths of the Atlantic ocean.

A nation can "sink" in many ways. It can lose several degrees of the magnitude it might once have possessed. It can degrade itself by sinking into moral morass. It can destroy itself by internal warfare, political strife, or technologically provoked cataclysm.

One need not claim that the fabled Atlantis existed on North American shores, but there is an astounding amount of evidence that a number of forgotten people and cultures did flourish in what is now the United States. Those scientists who somehow still cling to a kind of orderly evolutionary progression of Mongolian types transforming themselves into characteristic Amerindian types have got to be ignoring a great many discoveries of diverse and apparently anomalous genetic strains of *Homo sapiens* which once thrived on the North American land mass.

Few students of American history are aware that what is now the desolate Death Valley area was once a veritable Garden of Eden, complete with majestic palm trees and a proud people of heroic proportions.

In the June, 1970, issue of *Wild West* magazine, Ed Earl Repp told of the "honor and privilege" which was his in working with H. Flagler Cowden and his brother Charles C. Cowden, scientists "dedicated to the study of desert antiquity." Repp was present when the Cowden brothers uncovered the skeletal remains of a "human being believed to be the largest and oldest ever found in the United States."

It was in 1898, according to Repp, that the Cowdens discovered the human fossil remains of a giant female, "who was a member of the race of unprecedented large primitives which vanished from the face of the earth some 100,000 years ago." Although the scientists of that time did not have our modern dating methods, Repp states that the Cowdens were able to reach conclusions of time and age by the amount of silica in the soil and sands and by the state of petrification of the skeletal remains, along with the crystallization and opalization of the bone marrow.

"In the same earth strata where the giant female skeleton was found," Repp recalled, "they also recovered the remains of prehistoric camels and mammals of . . . an elephantlike creature with four tusks instead of the present-day two. With them were the remains of petrified palm trees, towering ferns and prehistoric fishlife."

The Amerindians were not an abnormally tall people by any means. The Shoshones, Paiutes, Cosos and other desert tribes that occupied the valley at the time of the European invasion of the continent would have been dwarfs in comparison with the unknown race of prehistoric giants. If we may assume the same kind of height ratio with which we are familiar between the sexes then we might suppose that the men of the vanished valley paradise must have been eight feet tall.

Neither Neanderthal nor Cro-Magnon were taller than *Homo sapiens*. Who were these mysterious Goliaths of Death Valley?

Repp tells us that in the same pit in which the Cowdens found the skeleton of the giant female they unearthed the petrified remains of marine life, indicating that Death Valley may have been an inlet of the Pacific Ocean at the time the lost race of giants lived there. That the people were tall is further indicated by the discovery of "handhewn caves high up in the chalklike cliffs, almost inaccessible from either top or bottom approaches."

Repp writes that the Cowdens discovered a number of anomalous physical appendages and attributes not found in contemporary man. There were ". . . the existence of several extra 'buttons' at the base of the spine . . . and every indication betraying the woman and her people were endowed with a tail-like appendage. In her jaws, the canine teeth were twice the size in length than modern man."

The Cowdens theorized that when the California which we know today was formed, together with the rising of mountains and the retreat of the seas, the tropical climate left the valley regions. The steaming swamps were replaced by vast wastelands, which still remain over much of the southern portion of the state. Then, with the advent of the Ice Age ". . . the freezing northern blasts swept down upon the tropical beasts and humans, who wore little or no clothing, and literally froze them to death in their tracks. . . . The glacial icepacks moved down into Death Valley . . . at the rate of 4½ miles per hour, burying all life beneath layer-upon-layer of petrifying silt and glacial mud."

The fossilized remains of the seven and one-half foot woman were found at a depth of five feet in a "hard-rock formation of conglomerate containing small amounts of silica, which required longer time to petrify than normal desert sands."

Perhaps prehistoric California was the home of the Amazons, those legendary, statuesque female warriors; for in July, 1895, a party of miners working near Bridlevale Falls found the tomb of a woman whose skeletal remains indicated that she had stood six feet, eight inches in height.

G. F. Martindale, who was in charge of the miners, noticed a pile of stones that seemed to have been placed against the wall of a cliff in an unnatural formation. Assuming the rock had been stacked by human hands, Martindale told his men to begin removing the stones in order to investigate what might lie beyond the formation.

The miners were astonished when they found a wall of rock that had been shaped and fitted together with apparent knowledge of masonry. The joints between the blocks were all of a uniform eighth-of-an-inch thickness, and a contemporary news account quoted one of the men as stating that the stonework was ". . . beautiful . . . as pretty as any wall on any building that I have ever seen."

The miners felt at first that they might have stumbled upon some lost treasure trove, and they set about tearing down the wall so that they might claim their found wealth.

Instead of riches beyond their wildest imaginings, however, the men found a large mummified corpse lying on a ledge that had been carved from the natural stone. The miners lighted their carbide head lamps and attempted to translate their disappointment into a more profitable examination of the burial vault, but all the chamber contained was the mummy of a very large woman. The corpse had been wrapped in animal skins and covered with a fine gray powder. She was clutching a child to her breast.

When the mummy was taken to Los Angeles, scientists there agreed that the woman was the citizen of a race that had thrived on this continent long before the American Indian had become dominant. They further arrived at a consensus that the woman's height of six feet, eight inches would have represented a height in life of at least seven feet. Figuring the classic height difference between men and women, they supposed that the males of the forgotten species would have been nearly eight feet tall.

A veritable catacomb of the skeletal remains of this lost race of giants was found when workmen were opening a way for the railroad between Wildon and Garrysburg, North Carolina. According to a contemporary newspaper account dated April 4, 1874, the bodies exhumed were of a "strange and remarkable formation."

> The skulls were nearly an inch in thickness; the teeth were filed sharp, as those of cannibals, enamel perfectly preserved; the bones were of wonderful strength, the femur being as long as the leg of an ordinary man, the stature of the body being probably as great as eight or nine feet. Near their heads were sharp stone arrows, some mortars . . . and the bowls of pipes, apparently of soft soapstone. The teeth of the skeletons are said to be as large as those of a horse.

The bodies were found closely packed together, laid tier on tier, as it seemed. There was no discernible ingress or egress to the mound. The mystery is, who these were, to what race they belonged, to what era, and how they came to be buried there. To these enquiries no answer has yet been made, and meantime the ruthless spade continues to cleave body and soul asunder, throwing up in mangled masses the bones of this heroic tribe. It is hoped that some effort will be made to preserve authentic and accurate accounts of these discoveries, and to throw some light, if possible, on the lost tribe whose bones are thus rudely disturbed from their sleep in earth's bosom.

Daily Independent, Helena, Montana

The *Dallas Morning News*, July 30, 1974, carried yet another account of the discovery of a seven-foot woman. Frank X. Tolbert stated that Dr. Ernest (Bull) Adams, an attorney in Somervell County and an amateur archaeologist, found her bones sealed in a cave at the crest of a high mesa near the hamlet of Chalk Mountain. The complete skeleton of the woman, which Dr. Adams discovered sometime in the middle 1950s, is now on display in a glass case in the Somervell County Museum on the courthouse square in Glen Rose, Texas.

Dr. Adams believed that the woman was of average size for her unknown race and that ". . . the cave was a maternity ward for these giants . . . the cave was steam-heated by water boiled under the floor . . . the woman had died in childbirth apparently. And her perfect teeth suggested she was quite young."

The New York Times for May 4, 1912, told of a find of several skeletons of gigantic humans made while excavating a mound at Lake Delavan, Wisconsin. News of the discovery was brought to Madison, Wisconsin, by Maurice Morrissey, who, in turn, informed the curator of the State Historical Museum. According to the newspaper account, eighteen skeletons were found in one large mound at a Lake Lawn farm:

The heads, presumably those of men, are much larger than the heads of any race which inhabit America today. From directly over the eyesockets, the head slopes straight back and the nasal bones protrude far above the cheekbones. The jawbones are long and pointed, bearing a minute resemblance to the head of a monkey. The teeth in front of the jaw are regular molars.

There were also found in the mounds the skeletons, presumably of women, which had smaller heads, but were similar in facial characteristics. The skeletons were embedded in charcoal and covered over with layers of baked clay to shed water from the sepulchre.

No sooner must we confront the question of the identity of this mysterious race of lost giants then we discover that the skeletal remains of people less than two feet tall have also been discovered on this continent. We may almost feel cheated that we must remain so totally ignorant of so wonderfully diverse a range of cultures and peoples who have flourished in this land in a world before our own.

Harper's Magazine, July, 1869, stated that Tennessee newspapers for the year 1828 told of several burying grounds, from a half acre to an acre in extent, which were discovered in Sparta, White County, Tennessee, in which extremely small people had been interred in tiny stone coffins. The tallest of the wee folk discovered was 19 inches.

Lest one think that those who discovered the strange burying grounds had merely found an infant cemetery for the giants, contemporary accounts describe the bones of the small ones to have been "strong and well set, and the whole frames well formed."

The graves were only dug about two feet deep, and the tiny corpses had been buried with their heads to the east, laid on their backs, with their hands folded across their chests. In the bend of the left arm of each skeleton lay a pint vessel made of ground stone or shell of a grayish color. Each vessel contained two or three shells. The skeletons were regular and uniform, with the exception of one that bore 94 pearl beads about its neck.

According to *Harper's Magazine*, a work published in 1853,

The Romance of Natural History, also refers to diminutive sarcophagi that were found in Kentucky and Tennessee.

Nearly everyone loves a mystery, but most people desire an eventual solution to prevent their being driven mad from frustration over the lack of any contributive clues. What once seemed a neat, progressive, evolutionary line has become hopelessly convoluted and chaotic. While many archaeologists and anthropologists are debating the origins of the Amerindian and the date of his arrival on this continent, few chose to pursue the erratic path dotted with the skeletal remains of giants—some of whom might even have had tails—and the evidence that this continent appears to have supported "worlds upon worlds" before any "Adam" of our epoch set foot on these shores.

New World? Bah! Humbug! Civilizations have flourished and have been devastated on this continent many times in an unknown past—perhaps before Egypt was more than a dream and long before Greece constructed her first city-state.

CHAPTER SEVEN

Mysterious Master Builders

Is it possible to determine the precise date of the end of the world by utilizing certain calculations derived from the Great Pyramid at Giza?

Is it true that the "pyramid inch" is equal to a year in prophecy?

Were the pyramids built by architects from Atlantis?

Or from some extraterrestrial source?

Is the Great Pyramid of Cheops the physical embodiment of a lost science of vast antiquity and unsurpassed knowledge?

Although everyone who has pursued the matter of pyramids to any degree is able to agree that the Great Pyramid is at least 4,000 years old, further assertions as to who built it, when, and why are certain to encounter opposing theories.

As to *what* the Great Pyramid is, Peter Tompkins in his *Secrets of the Great Pyramid* states that it has been established that it is ". . . a carefully located geodetic marker, or fixed landmark, on which the geography of the ancient world was brilliantly constructed; that it served as a celestial observatory from which maps and tables of the stellar hemisphere could be accurately drawn; and that it incorporates in its sides and

119

angles the means for creating a highly sophisticated map projection of the northern hemisphere. It is, in fact, a scale model of the hemisphere, correctly incorporating the geographical degree of latitude and longitude.''

Tompkins hypothesizes that the Great Pyramid may well be the "repository of an ancient and possibly universal system of weights and measures, the model for the most sensible system of linear and temporal measurements available on earth, based on the polar axis of rotation . . . whose accuracy is now confirmed by the mensuration of orbiting satellites.''

Until quite recently, there was no proof that the ancient Egyptians had any of their number capable of planning or constructing such a magnificent edifice as the Great Pyramid. But now, according to Tompkins, it is quite clear that whoever built the Great Pyramid "knew the precise circumference of the planet, and the length of the year to several decimals—data which were not rediscovered till the seventeenth century. Its architects may well have known the mean length of the earth's orbit round the sun, the specific density of the planet, the 26,000 year cycle of the equinoxes, the acceleration of gravity and the speed of light.''

Even today with our skyscrapers reaching for the clouds in all of our major cities, the Great Pyramid is still the world's most massive building. And only in this recent generation of man has the Great Pyramid been surpassed as the world's tallest building. The Empire State Building in New York City is among the very highest buildings ever erected by modern man, yet it is only about two-fifths the volume of the Great Pyramid.

The Great Pyramid is the only remaining "wonder" of the seven legendary marvels of the ancient world.* With the re-

---

*The others: the gardens of Semiramis at Babylon, the statue of the Olympian Zeus by Phidias, the temple of Artemis at Ephesus, the mausoleum at Halicarnassus, the Colossus of Rhodes, and the Pharos (lighthouse) at Alexandria. In some listings, the Walls of Babylon are substituted for the Alexandrian Pharos.

markable endurance which it displays, it may be the only wonder remaining after our epoch has had its play.

Who was the master builder who drew the plans for this architectural marvel? It would appear that the architect for this greatest of Egyptian pyramids was not even an Egyptian. His name spelled in Egyptian is Khufu. The Greeks called him Cheops. But according to the third-century Egyptian historian Manetho, Khufu was "of a different race."

The famous fifth-century Greek historian Herodotus states that the builders of the Great Pyramid were *shepherds*.

This seems all the stranger when Genesis tells us that to the Egyptians "every shepherd is an abomination." The Egyptians did not seem to have any sort of cowboy mystique, and they employed others to tend their flocks and herds. Yet according to many records, Khufu, or Cheops, the master builder of the Great Pyramid, was a shepherd.

Some Bible students will immediately recall that the Pharoahs who ruled during the Israelites' Egyptian sojourn set them to building pyramids. Since the Israelites were a nation of shepherds, could Khufu have been an Israelite architect who designed the Great Pyramid before Moses led the enslaved hosts on their exodus?

It would seem unlikely, for the pyramid at Giza was constructed much earlier than the Israelites' Egyptian Period, and the pyramids that the children of Israel tugged and heaved together are deemed to be rather shoddy and hastily erected duplicates of the Pyramid of Cheops.

In a paper published by Ambassador College, Herman L. Hoeh assembled some interesting data, together with some remarkable hypotheses, which may present a clearer portrait of the enigmatic architect of the Great Pyramid.

Cheops was not a polytheist, for Herodotus records that he closed the temples and prohibited the Egyptians from offering sacrifices. The deity whom Cheops served was named "Amen" in the ancient Egyptian spelling. Strange as it may seem, Hoeh reminds us, "one of the names of Jesus Christ, from the Hebrew, is 'Amen.' " (Revelation 3:14)

Joseph, the wise Israelite visionary and dream interpreter, who was sold into Egyptian captivity by his jealous brothers, rose to prominence under the Pharoah of Upper Egypt, who was named Amenemhet III. Because "Amen" appears to have been a common name among the Pharoahs in Joseph's day, Hoeh maintains that the rulers must have been strongly influenced by the religion of Cheops.

Hoeh finds additional fuel to fire his hypothesis of Creator-inspired Cheops in that Pharaoh Amenemhet gave Joseph "to wife Asenath the daughter of Potipherah priest of On" (Genesis 41:45). On is but another name for the god Amen, states Hoeh, and he goes on to point out that in Revelation 1:8, in the original, inspired text of this verse, "the Greek word Christ used was 'On'—the 'existing one.'!"

Hoeh's research convinces him that Cheops was a contemporary of King Zoser of Egypt (1737-1718 B.C.), who built the "step pyramid" a short time before Cheops constructed the Great Pyramid. Zoser ruled part of Lower Egypt while Joseph served as Prime Minister, under Amenemhet III, who was king of Upper Egypt and Pharaoh of all Egypt.

Egypt at this time appears to have been a confederation of powerful city-states ruled by lesser kings serving one pharaoh. Cheops was a foreign king of an Egyptian city-state, whose domain reached into the Delta of Egypt.

Hoeh believes the evidence is clear that Cheops must have constructed the Great Pyramid during the beginning of the Israelites' sojourn in Egypt (1726-1487 B.C.) and about the time of the seven years of famine. Hoeh further believes that the noted individual who assisted Cheops was none other than Joseph. History records the man as *Souf*, foreman of the works of Khufu, or as Saf-hotep, one of twelve brothers (Joseph had eleven brothers) who built the labyrinth of Ancient Egypt for Amenemhet III.

Hoeh further cites a corrupted Egyptian story of the later life of Khufu in which he summons an aged sage to his palace. The sage was said to have lived to the age of 110. Genesis records the death of Joseph at 110 years of age.

It is written in the ancient texts that Cheops/Khufu also wrote a major work of scriptural importance. Manetho, the Egyptian historian, wrote that Cheops was "arrogant toward the gods, but repented and wrote the Sacred Books . . . a work of great importance."

In answering the question of which sacred book, Hoeh does a bit of dazzling literary-historical detective work and names Job, author of the Book of Job, as none other than Cheops, builder of the Great Pyramid.

How did the central character of one of the most famous religious allegories in all of literature become the king of an ancient Egyptian city-state responsible for an architectural marvel that became one of the seven wonders of the ancient world?

Here, greatly encapsulated, is Hoeh's reasoning:

It seems likely that Cheops' sacred work was not an Egyptian book, since he closed the polytheistic temples and emphasized worship of one god. Cheops was a foreigner, impressing his religion, as well as his politics, upon his Egyptian subjects. They would not be likely to preserve the sacred book of a man whose religious views they would later oppose.

Cheops/Khufu had yet another name, Saaru of Shaaru. Shaaru is another designation for the inhabitants of the region of Mt. Seir. Cheops' domain extended from Mt. Seir to Lower Egypt at the time of Joseph.

"Mt. Seir was famous in history as the 'Land of Uz' . . . Uz was a descendant of Seir the Horite (Genesis 36:28)," writes Hoeh. "The Arabs preserve a corruption of Cheops of Mt. Seir or the Land of Uz. They call him the 'wizard of Oz.' "

In Hoeh's opinion, it all adds up:

"Now what individual who dwelled in Uz was arrogant, repented of his sin, and wrote a Sacred Book? None other than Job! And the Sacred Book is the *Book of Job* !"

As an added fillip, Hoeh informs us:

"The ancient Greeks called Job 'Cheops'—pronouncing the letters 'ch' almost as if they were an 'h.' We call Job 'Hiob' in German—and we pronounce the final 'b' as if it were a 'p' much as the Greeks did. Plainly, Cheops is but an altered pronunciation of Job!"

Hoeh offers numerous Bible verses to further substantiate his hypothesis, and concludes that Job/Cheops/Khufu built the Great Pyramid as a monument to "commemorate what Joseph did for Egypt and to mark the border of the territory given to Joseph's family in the land of Egypt by the Pharaoh."

While Hoeh builds a good case for his identification of Cheops as the god-tested Job of the Old Testament, one might build an equally good case for the supposition that Cheops might have been an "ancient astronaut" from some extraterrestrial world.

Cheops was said by the Egyptians to have been of a "different race" from them. Perhaps this "difference" might have been much more alien than simply being an Israelite.

Might Cheops have been called a "shepherd" because it was deemed that he had come to guide and to comfort the people of his earthly domain?

With his teaching of one god over Egypt's polytheistic heirarcy, did Cheops reveal his higher consciousness?

Perhaps the Sacred Book was a compilation of metaphysical and scientific secrets, the like of which earned him the title of the "wizard of Oz."

And might he have left the pyramid for posterity, not to commemorate any earthly event, but to serve as a beacon light shining through history to alert mankind to the evidence that they are not alone in the universe? The pyramid may exist as some cosmic educational toy, which serves as an irritant in the mind of modern man as he tries to ponder what appears to be an ancient science somehow out of context with the nations that surrounded Egypt over 4,000 years ago.

Whether Cheops/Khufu was an ancient astronaut or ancient Israelite, he provided an incredible number of generations with a tantalizing mystery of why he constructed such an edifice and for whom.

Sometimes even our already solved mysteries suddenly breed new wrinkles and ramifications.

For example, archaeologists felt that they had solved the question of how the Mayan farmers conducted the agriculture that had to support the heavy population of their city-states in the Yucatan peninsula of Mexico. In this case, they assumed that they had the present to illuminate the past.

Today's Mayan farmer employs the "milpa" or primitive "slash-and-burn" system of land usage. In this method, trees and jungle growth are burned away from prospective farmland. After the first rain, seed holes are drilled into the soft soil with a sharp stick. The land is cultivated for one more season, then the field is abandoned and a new clearing is burned in the jungle to provide land for the next crop.

Certain archaeologists have theorized that the slash-and-burn techniques of clearing farmland may have been one of the factors that depleted the Mayan's land and precipitated a decline of their civilization some time around 900 A.D.

But just a few years ago, archaeologists uncovered a canal-reservoir system on the Yucatan peninsula in Mexico that indicates that an early, pre-Mayan culture supported an advanced knowledge of hydraulic engineering and provided its people with extensive agricultural benefits more than 2,000 years ago. Some unknown culture established a network of 30 canals and 25 man-made, large-scale reservoirs approximately 30 miles southeast of what is now the little capital city of Campeche on the Gulf of Mexico.

The archaeological excavators are from Brigham Young

University, Provo, Utah, and it is their opinion, as they labor through the processes of mapping, surveying, excavating, and studying the site, that they have found what will be the "oldest and most intricate water-collection system in pre-Mayan history and possibly the only one of its kind in the New World."

Dr. Ray T. Matheny, associate professor of archaeology, credits the unknown canal builders as having been skilled engineers who were able to determine the difficult gradients that permitted rainwater to collect in canals and flow into waiting reservoirs.

Nelson Wadsworth, writing in *Science Digest*, March, 1974, quotes Dr. Matheny as stating: "What we are finding at [the ruins of] Edzna is a fairly large city of several hundred families organized on a high level of political authority and sustained by a stable economy which we believe was pot-water farming. This type of canal and rainwater collection for agriculture was simply not known elsewhere on the flat lands of the New World."

Dr. Matheny declares the site to be unique because of the deep soils which are rare on the Yucatan peninsula, together with the possible use of canals for pot-watering. Such intensive agricultural techniques are not generally known or practiced by the Mayan people of today.

The archaeologists found that rainwater from the adjacent jungle still drains into the huge feeder canal, then flows into a moat surrounding a fortress before it moves into a number of nearby reservoirs. The team from Brigham Young measured the canal as 12 kilometers long, and, in some areas, 50 meters wide.

Dr. Matheny commented: "This in itself was a tremendous construction achievement. It could not have been built without some kind of advanced engineering know-how and sophisticated social organization."

Although there is no mystery about who constructed the

170,000 miles of underground channels in Iran more than 3,000 years ago, the system of subterranean aqueducts is an awesome achievement that still provides 75 percent of the water used in that country today.

With the exception of areas in the northwestern provinces and the southern shores of the Caspian Sea, Iran receives only six to ten inches of rainfall each year. But far from being a barren, scorched region, Iran has been a self-sufficient agricultural nation for centuries. It has been able to grow its own food, and it has managed to produce crops for export by tapping underground water through the means of an ingenious system called *qanat* (from a Semitic word, "to dig").

H. E. Wulff explained the system in the April, 1968, issue of *Scientific American*:

"The qanat system consists of underground channels that convey water from aquifers in highlands to the surface at lower levels by gravity. . . . The qanat works of Iran were built on a scale that rivaled the great aqueducts of the Roman Empire. Whereas the Roman aqueducts now are only historical curiosity, the Iranian system is still in use after 3,000 years and has continually been expanded. . . ."

The qanat shafts may go as deep as 300 feet to reach water, and such depths require windlasses at intervals of 100 feet apart. Guide shafts are sunk at intervals of about 300 yards to indicate the route and the pitch of the conduit for the well diggers. The mouth end of the conduit is lined with stones to protect it from storm-water damage. The tunnels and shafts need no reinforcement when they have been dug through hard clay or a coarse conglomerate that is well-packed.

"Not until the qanat has been completed and has operated for some time is it possible to determine whether it will be a continuous 'runner' or a seasonal source that provides water only in the spring or after heavy rains," writes Wulff.

The initial investment in constructing a qanat is considerable, and the owners and builders will often resort to laborious devices in order to insure or enlarge its yield. Careful attention is also given to maintenance of the qanat.

"As is to be expected of a system that has existed for thousands of years and is so important to the life of the nation, the building of qanats and the distribution of water are ruled by laws and common understandings that are hallowed by tradition," Wulff states.

"The builders of a qanat must obtain the consent of the owners of the land it will cross, but permission cannot be refused arbitrarily. It must be granted if the new qanat will not interfere with the yield from an existing qanat, which usually means that the distance between the two must be several hundred yards, depending on the geological formations involved."

As regards the mysterious walls of the Berkeley and Oakland hills in California, no one seems to have the foggiest idea who built them so that their line of progression is not at all "ruled by laws and common understandings that are hallowed by tradition."

The rather ordinary-looking stone walls are found mainly in heavily wooded or chaparral-covered areas. Although in a few places it appears as if they might have been utilized as some kind of fortification, they seem to fulfill none of the usual functions of walls.

Sibley S. Morrill investigated the matter of the perplexing walls for the October, 1972, issue of *Pursuit*. According to his research, the walls survive in sections ranging from 20 feet to more than 200 yards in length. In height, they vary from two feet or less to five feet or a bit more. Their breadth at ground level is a rather impressive four feet. In Morrill's opinion, such breadth makes it ". . . a near certainty that the walls originally were much higher through the use of smaller stones along the top."

Limited digging near the base indicates that the rock goes down about ten inches below the surface. It is difficult to imagine the wall serving any nonfunctional purpose, since some of the rocks employed in their construction weigh more than 200 pounds.

A Mr. Seth Simpson of Oakland is said to have studied the walls as a hobby for a good many years. According to Simpson, the walls extend for nearly seven miles south into the Oakland hills, but he has not been able to relate them to any known boundary markings:

"Water company survey maps show that none of the walls has any detectable relationship to boundary lines; except for one case in the Vollmer Peak area, boundary lines parallel no walls nearer than about 600 yards."

Neither does any clue remain to suggest that they might be the remains of any sort of animal pen or corral.

"They are, for the most part, straight," Morrill informs us. "Some intersect at an angle, and there are instances of parallel walls separated by as much as ten yards or so, but there are no indications whatever that they formed enclosures."

In the hills behind Milpitas, Simpson discovered similar walls in a gently rolling, almost treeless country.

Again, with only one exception, "they offer no suggestion of the usual purpose for a wall. . . . The walls are virtually useless as fences. . . . In the Milpitas area, the stone walls just run their way for a few score or few hundred yards and then stop."

No area resident appears to know anything about the origins of the walls. The walls have just always been there.

An examination of old newspaper clippings concerning the reactions of various residents of the area to the builders' walls reveals opinions ranging from Mexican to Chinese, from pioneers to "Mazatlánes," sounding "strangely similar to Atlanteans, to whom the Aztecs and their predecessors, who lived about Mazatlán, down the Mexican coast, were reputed to be related."

Seth Simpson concedes that it is possible that at least some

129

of the walls might have been built by Indians for the purpose of driving game into "a sort of cul-de-sac where they could be easily killed."

Simpson also states that, in the absence of extensive excavations along the walls, such a theory is all he can suggest: "They were built by unknown persons, in an unknown year, for an unknown purpose. And very possibly, despite our hopes, they will remain a puzzle for an indefinite future."

The London *Daily Telegraph* for January 27, 1973, provided the details of another puzzling wall that appeared to go nowhere for no particular reason.

The recorder of rhis mystery wall was the eminent Sir Alec Kirkbride, who, at age 75, was identified by the *Daily Telegraph* as one of the last surviving British officers to have fought in the Arab revolt with the famous Lawrence of Arabia. Kirkbride's wall runs for 20 miles at a distance of about 12 miles from Petra to Jordan.

Sir Alec, who was for many years British Resident in Transjordan and terminated his long career as British Ambassador in Libya, informed the press that he first spotted the "curious wall" when he flew over it in a light airplane. He was so intrigued that he returned to take a closer look from horseback.

"It was utterly staggering," Sir Alec recalled, "because it had involved a tremendous amount of labor, being about ten feet wide and two feet tall. But it bore no relation to any boundary or defensive position at all. It's just a great jumble."

Sir Alec's theory is that the wall was built by the Nabateans in the early Christian era, but he admitted that he is far from convinced that those desert peoples really did construct the mystery wall. "Despite much research, no one can explain it," he commented.

In the April, 1973, issue of *Pursuit*, a staff writer offered the following observation regarding Sir Alec's desert wall to no-where:

". . . Sir Alec, who certainly knows the area as well as anyone can, states flatly that this wall is not a defensive or boundary wall. In fact, its dimensions make it sound more like a road; but the description suggests that it doesn't go any-where. If all this is so, and our figures are correct, it means that some group of (presumably demented) people carefully piled up 2,112,000 cubic feet of rock for no reason whatso-ever. . . . One cannot seriously entertain the idea that people lugged stones around for the fun of it, particularly in the in-hospitable area around Petra."

CHAPTER EIGHT

Henges, Observatories, and Computers

For over half a century the fragment of carved reindeer bone rested in a darkened corner of a French museum, ignored, never closely examined. When Alexander Marshack slipped it under his microscope on that July day in 1965, he suddenly felt that he "was looking into the mind of the man who had fashioned and used it more than thirty thousand years ago."

The serpentine series of marks that had been etched on the surface of the bone had taken on new meaning. An intentional pattern had become obvious.. Marshack believed that he had unlocked the puzzle when he interpreted the artifact, formed during the late Ice Age, as bearing notches coincident with phases of the moon over a period of more than two months.

"Whether or not the image is accepted as lunar," Marshack stated in the January, 1975 issue of *National Geographic*, "I had discovered the earliest known human notation, made more than twenty thousand years before the development of writing, arithmetic, or the calendars of later cultures we regard as civilized."

Although Marshack's opinions concerning Cro-Magnon man are controversial, he is convinced that far from being little more than a primitive toolmaker and hunter, Ice Age man was ". . . a more modern human, thinking in a far more sophisticated way than was believed possible."

To aid him in his research, Marshack employed extreme closeup photography together with a geologist's field microscope to magnify the engraved notations and compositions of Cro-Magnon man. In his analyses of various cave paintings, he used ultraviolet and infrared photography, "literally to shed new light on the surprisingly rich art of Cro-Magnon man."

It seems as if from the very advent of his intellectual history, man has been extremely concerned with marking and measuring the ebb and flow of his life, the microcosm, and the reach and influence of the stars and planets, the macrocosm.

Perhaps the greatest *wonder* we experience in regard to the origins of man's recording of time is how extremely sophisticated the ancients were in their calibrations of the movements of the heavenly bodies.

Perhaps the most controversial *question* is whether these so-called "primitive" cultures developed these techniques on their own, or if they inherited the methods from an earlier, advanced civilization whose precise identity remains uncertain to us today.

Twenty-five-hundred years ago, the Babylonian astronomer Kidinnu determined the value of the annual movement of the sun and moon with an accuracy that was not bettered until 1857, when Peter Andreas Hansen, director of the Gotha observatory, attained figures with not more than three seconds of arc in error. The figures of the ancient Babylonian had an error of less than nine seconds of arc.

Kidinnu's estimations of lunar eclipses were two-tenths of a second from the measurement established by our most modern methods of calculation.

In 1887, Theodore von Oppolzer established the present system of calculation. His estimates included an error of seven-tenths of a second in arc in estimating the movement of the sun.

Just how Kidinnu, conducting his research in 500 B.C., could have arrived at such accurate figures without the use of telescopes, precision-timing devices, or a high degree of mathematical sophistication is rather astonishing.

In a series of lectures at Great Britain's Leeds University in 1957, Toulmin and Goodfield suggested that the Mesopotamian astronomer may have had access to a body of astronomical knowledge that extended back into "an unknown period of scientific inquiry" that may have been amassed over a much longer period than the background available to our modern astronomers.

In a report to the United Nations in 1951, Professor Tung-Stso-pin stated that the Shang dynasty (1700–1100 B.C.) utilized a lunar-solar calendar, the *Ssu-Fen*, whose months consisted of 29 or 30 days, with a precise length of 29.5305106 days. Our modern calendars are more accurate by only eight hundred-thousandths.

Chang Heng, an astronomer who lived between 78–139 A.D., asserted that the Earth was egg-shaped and that its axis pointed toward the Polar star.

The ancient book *Huang Ti-Ping King su Wen* contains the statement by the Chinese scholar Chi-Po (circa 2697–2596 B.C.) that Earth floats in space.

An oracle bone found at Anyang revealed an inscription telling of an eclipse of the Moon "on the 15th day of the 12th moon of the 29th year of King Wu-Ting"—i.e., November 23, 1311 B.C. The inscription goes on to record that the Chou ruler, Chou-wen-wang, had ordered a sacrifice in 1137 B.C. to establish a cosmic balance after an eclipse occurred *one day* later than predicted. It is astonishing to discover that the Chinese astronomers of over 3,000 years ago could predict lunar eclipses to such a degree of accuracy.

Contemporary scholars, who have analyzed the Orphic Hymns, which date back a thousand or more years before the great poet Homer, have discovered firm evidence that the ancient Greeks understood that the seasons were caused by the Earth's rotation around the sun along the ecliptic. They also possessed knowledge of the equinoxes and solstices and appeared to know that the seeming rotation of the stars was due to Earth's revolving on its axis. The Orphics also refer to the presence of mountains on the Moon, a fact that would not be firmly established in Western science for another two thousand years.

In his paper, *Measuring the Macrocosm*, J. R. Jochmans assesses the astronomical knowledge of the ancient Greeks by citing such quotations from their scholars and scientists as the following:

Permenides (sixth century B.C.): "The Moon illuminates the nights with borrowed light."

Empedocles (fifth century B.C.): "The Moon circles about the Earth with borrowed light."

Democritus (writing twenty centuries before the telescope): "The markings on the Moon . . . are shadows from high mountains and deep valleys."

Anaxagoras (recording his thought 2,500 years ago): "It is the Moon that darkens the Sun during an eclipse."

Anaxagoras also asserted that it is the Earth's shadow that falls on the Moon during a lunar eclipse. And he wrote about "other Earths" with inhabitants.

Posidonius (135–50 B.C.) conducted a study of the tides and concluded that they were associated with the movements of the Moon around Earth. When the German astronomer Kepler arrived at the same conclusion only 350 years ago, he was severely criticized for his research assessment.

Thales of Miletus (fifth century B.C.) stated that the stars were made of the same substance as Earth. The concept of the universality of matter vanished in the Middle Ages with its egocentricity of perceiving Earth as the center of the universe. Such thought was not resurrected until the early twentieth century.

Heraclitus, together with the disciples of *Pythagoras* (sixth century B.C.), maintained that each star was the center of a planetary system.

Metrodotus (third century B.C.) argued for a multiplicity of inhabited worlds. He insisted that to believe the Earth to be the only inhabited planet would be as unwise as to believe that there was only one seed of grain growing in a large field.

The early Greeks obviously enjoyed a high degree of intellectual freedom. In the year 1600, the scholarly Dominican monk, Giordano Bruno, was burned at the stake for expostulating the heretical thesis that there were an infinite number of suns with planets revolving around them, which might be capable of supporting life.

Aristarchus (third century B.C.) observed: "Earth revolves in an oblique circle around the sun while it rotates at the same time about its own axis." Kepler was not to demonstrate this precept for another 1,500 years.

After one has examined the above statements and observations of the ancient Greeks, the discovery of a Greek calendar computer circa 80 B.C. made by Dr. Derek De Solla Price, Avalon Professor of History, Yale University, becomes almost expected, rather than astonishing.

The mechanism was originally brought into the twentieth century by a group of sponge-fishers, who discovered it, together with a pile of bronze and marble statues, in the wreck of an ancient Greek ship off the islet of Antikythera shortly before Easter of 1900. In his "Gears from the Greeks," published in the *Transactions* of the American Philosophical Society, November, 1974, Dr. De Solla Price writes that early assessment of the strange artifacts held that it might have been some sort of navigation equipment. His own evaluation establishes the mechanism as "a calendrical Sun and Moon computing mechanism which may have been made about 87 B.C."

According to Dr. De Solla Price's analysis:

> Perhaps the most spectacular aspect of the mechanism is that it incorporates the very sophisticated device of a differential gear assembly for taking the difference between two rotations, and one must now suppose that such complex gearing is more typical of the level of Greco-Roman mechanical proficiency than has been thought on the basis of merely textual evidence. Thus this singular artifact, the oldest existing relic of scientific technology, and the only complicated mechanical device we have from antiquity, quite changes our ideas about the Greeks and makes visible a more continuous historical evolution of one of the most important main lines that lead to our modern civilization.

Dr. De Solla Price states that the complexity of the device and its mechanical sophistication put it so far beyond the perimeters of what has formerly been assumed to be the limits of Greek technology that certain theorists have believed that it could only have come from alien astronauts arriving from outer space.

Although he sympathizes with the shock that one may experience at "revising upwards the estimation of Hellenistic technology," Dr. De Solla Price does not feel so radical an interpretation is necessary. Rather, he maintains, it makes more sense to admit that we have drastically underestimated the "whole story of Greek science."

The Antikythera device is ". . . an elegant demonstration or simulation of the heavens, more like an astrolabe perhaps than a direct ancestor of the calculating machines of Pascal and Leibniz. Nevertheless, it does use fixed gear ratios to make these calculations of the soli-lunar calendar, and it does this more by using pointer readings on a digital dial than by causing a direct geometrical modeling of the paths of the planets in space. The mechanism displays the cyclical sequence of sets of discrete phenomena, rather than a continuum of events in a flowing time. . . ."

Derek De Solla Price also restored the Tower of the Winds in Athens, in which an ingenious water clock once served as the timepiece for the Greek marketplace of 50 B.C.

A Macedonian astronomer, Andronikos of Kyrrhos, fashioned the clock, which probably included such elements as "an elevated reservoir constantly filled with water; beneath it a pot or tank into which water would drip at a carefully controlled rate; a float-chain-and-weight mechanism that turned an axle as the water level rose in the pot; and an astronomical 'clock-face' that revolved like a wheel at the end of the axle." The clock would have no hands as ". . . the position of the sun on the slowly revolving dial would tell the time of day."

Two generations before the birth of Jesus of Nazareth, the tower dominated the marketplace in the age of Caesar, Cleopatra, Mark Antony, and Octavius.

As De Solla Price visualizes the scene, the Tower of the Winds provided the visitor with an "unforgettably striking spectacle":

> . . . Within the circular railing at center, set off by fountains . . . turned a bronze disk—a model of the universe moving in harmony with reality. Among Andromeda, Perseus, and the figures of the zodiac, a golden sun, pegged into the proper hole for the time of year, moved behind a wire grid which indicated the hours of day and night and the lines of the horizon and meridian.
> . . . Truly the chamber was a place to deepen the thoughts of men. . . . This was no simple timepiece. It was a show-

piece, a symbol of intellectual triumph that proclaimed man's mastery of science. . . .

Later, during the Middle Ages, when that science was at its lowest ebb in Europe, the sophisticated citizens of the city-states of Central and South America displayed a knowledge of astronomy that appears to modern man to present an annoying enigma because of its remarkable accuracy.

A Mayan inscription on the wall of the tomb in the Pyramid at Palenque speaks of a month of 29.53086 days.

The Mayans at Copan estimated the lunar month as a bit less at 29.53020 days.

Contemporary astronomical measurement asserts that the correct mean figure is found in the average *between* the Palenque and Copan calculations.

The Mayans estimated the earthly year at 365.2420 days. Modern science has bettered that estimate by only two one-hundred-thousandths of a day more.

The Mayans were aware of the number of years and days it takes Jupiter to circle once around the zodiac.

They knew that Mars became a brilliant midnight star every 780 days.

They calculated the Venusian year and recognized the pattern of days that Venus appears as an evening star, and they were aware that the pattern was repeated each eight years.

The first astronomical conference on the North American continent took place 1,400 years ago in the ancient Mayan city of Copan, which is located deep in the jungles of what is now Honduras. The evidence of this noteworthy occasion was commemorated in an elaborately carved altarpiece, which has representations of the sixteen scientists in attendance on its side.

Writing in the February, 1974, issue of *Science Digest*, Douglas Colligan explains the Mayans' expertise in astronomy as having grown out of their "obsession with time and what they could learn from it. It was their belief that history would literally repeat itself, and by dating and studying past events they could anticipate what the future would bring.''

The people's concern over the future was both practical and religious. It was equally important to them to know when was the best time to plant crops as well as to know the best day to make a sacrifice to their patron god. The priest-astronomers were perceptive enough to realize that the steady movements of the planets constituted the most dependable timepiece they could find.

"The Maya wished to know which gods would be marching together on any given day because with that information they could gauge the combined influences of all the marchers, offsetting the bad ones with the good in an involved computation of the fates and astrological factors,'' Mayan scholar J. Eric S. Thompson wrote in his *Rise and Fall of Maya Civilization*. "On a successful solution depended the fate of mankind.''

Dr. Charles Smiley, chairman of the department of astronomy, Brown University, director of the Ladd Observatory, the leading expert on Mayan astronomy, pointed out to Douglas Colligan that part of the Mayans' studies were concentrated on what we today would call astrology—"trying to predict the future on the basis of what's happened in the past.''

Dr. Smiley went on to explain how the Mayan astronomers combined a bit of showmanship with their remarkable astronomical knowledge:

"Some people have insisted that the Mayans would predict solar eclipses and then make 'white magic' to prevent them. If they did, they were extraordinarily clever because if you predict an eclipse, make white magic and the eclipse fails to appear, you've been successful as a magician. If you predict an eclipse, make white magic and it happens, you've been a failure as a magician, but a success as a predictor.''

Although we have the evidence that the Mayans were able

to arrive at exact, or near exact, figures of the more important astronomical events, they possessed nothing like a telescope or any elaborate kind of astronomical equipment. The Mayan astronomers had to depend on what they could perceive with their technologically unassisted eyes.

Dr. Smiley speculates that the Mayans may have built their pyramids solely for the reason of keeping the heavy growth of tropical trees in check.

"They wanted to get up higher so they could look over the tops of the trees and see where the sun was rising, where the moon was rising," he commented. "I'm almost certain that the pyramids with distant markers provided their astronomical basis."

And it was from the tops of their pyramids that the Mayans were able to observe the heavens in such a meticulous manner that they could devise what Dr. Smiley calls a "solar eclipse warning table," which "not only warns of eclipses visible in Central America but all over the world."

Whatever genius devised the solar-eclipse warning table fashioned a technique that works just as accurately in the twentieth century and will remain accurate through the twenty-fifth century—just as far into the future as our own contemporary astronomers have plotted eclipses.

Dr. Smiley stated that although today's astronomers do not need to rely on Mayan techniques to predict future eclipses, the Mayans' habit of maintaining meticulous records of celestial events has much to offer the twentieth-century scientist.

"We are in the interesting position that if only we know a little more about the Mayan language, we can improve our astronomical tables," Dr. Smiley said.

Certain authorities on Mayan culture have theorized that the Caracol Tower at Chichén Itzá served the pre-Conquest astronomers as an observatory. In their report in *Science*, June 6, 1975, Anthony F. Aveni, Sharon L. Gibbs, and Horst Hartung state:

While we propose no grand cosmic scheme for the astronomical design of the Caracol, it can be inferred that the building, apart from being a monument related to Quetzalcoatl, was erected primarily for the purpose of embodying in its architecture certain significant astronomical event alignments, in the same sense that a modern astronomical ephemeris exhibits information of importance to us in the keeping of the current calendar. . . .

The directions of the faces of the Lower and Upper platforms of the Caracol seem to have been laid out deliberately to point to horizon events involving the sun and the planet Venus. Of the lines taken through the windows, the Venus setting points seem most plausible to us in view of both the accuracy with which they fit the architecture and the historical evidence bearing upon the importance of Venus to the Mesoamerican people. . . .

An interesting aspect of the Caracol Tower is that it looks for all the world like the ruins of a modern observatory. It is circular, constituting a new design element of its period, and it has a spiral staircase. The limestone masonry of the tower is set upon an elaborate platform substructure. The four lower-room doorways are aligned with the cardinal points, and the tower proper encloses an inner, circular chamber, with its four doorways set in alignment halfway between the first four.

Mayan authority J. Eric S. Thompson has observed that if one can free himself from the enchantment which antiquity sometimes induces and objectively observe the tower "in all its horror from a strictly esthetic point of view," he will find nothing quite so hideous as the Caracol. "It stands," he remarks, "like a two-decker wedding cake on the square carton in which it came. Something was pretty clearly wrong with the taste of the architects who built it."

But the tower may have been constructed solely for utilitarian, rather than esthetic, purposes. And certain scholars, such as John P. Molloy, University of Arizona, suggest that the astronomers of Caracol may have established a regional network of observatories that may have extended throughout significant sites in Central and North America.

"If correct, Molloy's theory would establish a very important case for intercultural collaboration in pre-Columbian astronomical observation," noted Robert D. Hicks, III, writing in the June, 1976, issue of *Sky and Telescope*.

Warren L. Wittry has described an astronomical observing station at Cahokia (circa 1000 A.D.), a large Amerindian settlement located in what is now St. Louis, Missouri. Called the American "Woodhenge" because of its superficial resemblance to the European henge stone monuments, the largest circle consisted of 48 wooden posts, the arc between each of them being seven and one-half degrees, while an additional post was approximately five feet east of the true center.

"From this post, one would have been able to look due east in line with a post on the circle and witness sunrise at the equinoxes," Hicks comments in his article. "Furthermore, from this same off-center post one could look past the fourth post north or south of due east to observe sunrise at the summer and winter solstices, respectively."

Remains of several villages of Wichita Indians (circa 1500 A.D.) have yielded "council" circles, each containing a low central mound surrounded by a ditch or a series of depressions arranged in elliptical patterns.

"The ellipses are located within sight of one another," Hicks reports. "The two at the Tobias and Paint Creek sites [Great Plains of Kansas] have their long axes aligned with summer solstice sunrise and winter solstice sunset. The Thompson and Hayes circles have their long axes in line with each other, as well as with winter solstice sunrise and summer solstice sunset. The motivation for these configurations was largely ritualistic, since ceremonial activities at the solstice were important to the people of this area."

Jonathan Reyman of Illinois State University has investigated astronomical alignments at various sites in the area of Chaco Canyon, New Mexico. A number of solstitial markers have been tentatively identified by the astroarchaeologist.

John P. Molloy has searched for astronomically significant alignments at the Big House, Casa Grande National Monument, near Coolidge, Arizona. This area was inhabited by the Hohokam Indians from 200 B.C. to 1475 A.D. The Big House itself was probably in use from 900–1100 A.D.

"[The Big House] is a three-story rectangular building of caliche and adobe, set upon a one-story platform substructure," Hicks writes. "The symmetrical plan contains five contiguous room tiers, four of which rise two stories and surround a central three-story tier. Astronomically significant alignments are found for windows and holes in the second-story west wall and for the third-story walls."

The various alignment holes all give lines of sight to the horizon and give the appearance of having been obviously placed with great care.

"Of the 14 openings investigated, eight were found to be celestially aligned, basically to the sun and moon, with the solstices and equinoxes indicated," Hicks explains. "Two sets of openings, Holes 6 and 7, are located directly above 8 and 9, all giving eastern horizon positions. This redundancy provided for 'fine tuning.'"

Molloy's research appears to have demonstrated numerous affinities between the Big House in Arizona and the Caracol Tower "observatory" in Yucatan, Mexico.

Hicks concludes by observing that such studies will undoubtedly provide "new insights into the ancient American past" and demonstrate that "pre-Columbian astronomy became well established for reasons that are only partly understood."

Significant intellectual breakthroughs often take time. For generations Stonehenge's 97-foot ring of 25-ton upright stones, topped by massive horizontal stone slabs, was considered little more than a remarkable achievement of late Stone Age and early Bronze Age men (between 1900 and 1600 B.C.) and later, a rather profitable tourist attraction. Scholars suggested that the henge had served as a mortuary, a crematorium, a sacrificial altar for perverse rites, and the scene of Druidic religious pageants.

While at one time or another Stonehenge might have served as all of the above, in 1965 Boston University Astronomy Professor Gerald Hawkins demonstrated that the gigantic marker stones comprised a 3,500-year-old observatory.

Hawkins fed the data of Stonehenge's positions of center point, archways, holes, and mounds into a computer and found that many of Stonehenge's alignments accurately pointed to the summer and winter solstice positions of the rising and setting Sun and Moon—the extreme north and south latitudes reached only on midsummer day and on midwinter day, the shortest day in the year. Hawkins concluded that Stonehenge was "locked to the Sun and Moon as tightly as the tides. It was an astronomical laboratory. And a good one, too."

Hawkins' subsequent research led him to assert that the astronomers of Stonehenge were so advanced that they had detected a phenomenon unnoted by modern astronomers—that eclipses of the moon occur in 56-year cycles.

According to *Time* magazine's summary of Hawkins' research (November 12, 1965):

> Hawkins, who inadvertently rediscovered the cycle after running Stonehenge eclipse data through a computer, immediately associated it with a mysterious circle of 56 "Aubrey" holes that ring the massive arches. He concluded that the holes formed a primitive eclipse computer. By placing a stone in each of six appropriate holes and moving them at appropriate times one hole around the circle, he decided, the Stonehenge astronomers had probably been able to tell accurately the dates when solar and lunar eclipses were apt to take place.

Since Hawkins' bold assertion that Stonehenge was more than an ancient Pagan altarsite, a number of henges have been identified in Great Britain and Europe, such as Callanish in Scotland and Newgrange in Ireland.

In October of 1973, Israeli environmental archaeologist Yehoshua Itshaki and other scientists announced that they may have found a Middle East precursor to Stonehenge at Rujum Al-Hiri. And according to the Israeli archaeologists, the scars of the five giant rings in the earth of the windswept moor indicate that Rujum Al-Hiri served as a combination calendar, direction finder, eclipse and star calculator before the pyramids were constructed and at least 1,000 years before Stonehenge was erected.

"Earth's axis is now oriented toward Stella Polaris instead of some other star," Itshaki stated. "So there are inconsistencies if you try to fix the position of stars today on the basis of what these people used over 5,000 years ago."

In other respects, though, Israeli scientists confirm that the calculator has maintained its accuracy down through the ages. For example, straight lines drawn through the centers of the overlapping circles will still lead one to Earth's true north. Itshaki emphasized that since basaltic rock of the Golan Heights resists the standard magnetic compass, the architects of Rujum Al-Hiri must have commanded considerable engineering skill to accomplish what they did.

Itshaki admitted that one of the facts that science will probably never know exactly is how long it took the creators of the Mideast Stonehenge to finish their project. A crude estimate has set the manpower investment at 2.5 million workdays.

Such an estimate develops at once into another mystery: Where did such an enormously large work crew live? There has not yet been found any evidence within miles of the astronomical calculator of any sizable settlement nearly as old as the giant rings.

Circumstantial evidence suggests that Rujum Al-Hiri's secrets of time and the planetary movements were passed on to succeeding generations by means of an oral tradition. Itshaki referred to the Jewish lunar calendar and its various feast days, nearly all of which are based on seasonal dates for the planting and harvesting of crops.

In commenting about the implications of Rujum Al-Hiri, It-shaki said: "What it might show is that this part of the world, halfway between the oases of Egypt and Mesopotamia, had organized society with ritual patterns and life styles much more advanced than we've previously had reason to believe."

An irritating enigma that so often remains in such discoveries as the one at Rujum Al-Hiri is that no one really knows *what* civilization was served by the combination calendar, direction finder, eclipse and star calculator.

CHAPTER NINE

The Skillful Hands of Unknown Artists

and Forgotten Physicians

I have been an admirer of comic-strip art since I was a boy in the 1940s. Many "comics" are not funny at all, but deal with serious representations of the human condition. Other strips are melodramatic and offer a "soap opera" interpretation of life. Escapist fare is available to those who follow the various "super heroes" in their nonending fight for law and order against the forces of corruption and organized crime. And there are always those delightful helpings of pure fantasy in which engagingly drawn animals are granted the power of witty and insightful speech, the prerogative of wearing clothing, and the dubious honor of enduring the human condition through the anthropomorphic transformation wrought upon them by the artist.

Not long ago, I came across three entertaining comic strips which, while new to me, I learned had been in circulation for quite some time.

The first, "Tutmouse, Pharaoh of the Mice," depicted the arrogant rodent seated upon his throne, fanning himself. Behind him, a mouse slave worked a much larger fan so that his excellency might be kept cool while he munched his meal of pumpkins.

Conflict, always a necessary part of any good story, whether humorous or serious, was represented in the appearance of a feline diplomat.

Resolution appeared imminent, however, since the eternal and implacable enemy of the whole of mousekind had come to offer Tutmouse an olive branch, a symbol of peace.

"The Friendly Fox Brothers" was the second comic strip I encountered on that day of discovery. This strip had to do with two foxes who turn vegetarian and obtain employment as herdsmen for a wealthy goat rancher.

Not only can the clever fox brothers utilize their knowledge of predators' ways to protect their new charges, but one of the pair is a talented maestro of the ancient wind instrument commonly known as a panpipe. If they cannot frighten or pummel away those who would do violence to the goat herds, Reynard will provide music to soothe the savage breast of the most vicious beast.

Obviously tales of herdsmen were very big in the area in which these comic strips were distributed, for the third set of panels dealt with the adventures of a cat that shepherds a flock of geese.

It would also seem likely that tales of conversion experiences and inspiring parables of redemption must also have been popular, since once again the comic strip offered a former predator now responsible for the very creatures he would once have deemed candidates for his meals.

The comic strips to which I have been referring were drawn on papyrus more than 3,000 years ago. I have allowed my own imagination to provide the names of the characters and the titles of the cartoon panels, but my thumbnail descriptions of the action contained within the drawings, though slightly exaggerated, is basically accurate.

Wilfrid D. Hambly, Curator of African Ethnology (retired), Chicago Natural History Museum, remarked in *Scientific Monthly* (October, 1954) that so far as humor in the sketching of animals is concerned, the efforts of Walt Disney and other cartoonists were anticipated by 3,000 years. And the unknown Egyptian artist based his animal antics on a basic principle that is followed today in animated cartoons and in comic strips: The humor is to be found in the dramatic contrast between the animals' real-life habits and their activities in the drawings.

It is not possible to determine precisely when man first made fire, when man first wrapped an animal skin around his body for warmth, or when he first picked up a piece of charcoal and began to draw on the wall of a cave. Some of the most realistic early drawings of wild animals were done in vivid colors on the walls of caves in France and Spain in the period of the Old Stone Age, but man-made carvings said to be twice as old as those famous cave drawings were found in Germany in the late 1950s.

Professor Walter Matthes, head of the College for Prehistorical and Early History Study in Hamburg, discovered the objects on a steep stone bank of the River Elbe. Professor Matthes stated his assessment that the carvings represent "the oldest man-made likenesses yet discovered" and estimated that the pieces were as much as 200,000 years old.

For the most part, the carvings are no larger than match boxes, and they depict the heads of human beings and Ice Age animals. According to Professor Matthes, the human heads bear few, if any, of the apelike characteristics so commonly associated with Neanderthal man.

It may upset the orderly scheme of things that some scholars have designed when objects of art are found that date back hundreds of thousands of years. Some scientists choose to believe that Neanderthal was an insensitive brutelike hulk of primitive *homo*, totally incapable of appreciating the aesthetic qualities of life around him—and certainly unable to fan any spark of creativity into a large enough flame to fashion paintings and carvings.

But few scientists seriously dispute the fact that Neanderthal man did live in Europe over 100,000 years ago. It is when elaborate works of art far older than man as a species is supposed to have existed are found in North America—where man was not supposed to have been at all until 20,000 or so years ago—that the perimeters of restrictive orthodox science are stretched far beyond their normal tolerance.

In March, 1891, a Mr. J. H. Hooper noticed what appeared to be a headstone to a grave on a wooded ridge on his farm in Bradley County, Tennessee. Naturally curious, Mr. Hooper dug around the stone, expecting to discover a name, together with the standard "rest-in-peace," born and deceased information. Instead, he found only a curious pattern of unknown characters in an undecipherable language.

In a report for *Transactions* of the New York Academy of Sciences, 11: 26–29, 1891, A. L. Rawson provided the following details:

> He dug deeper and uncovered other stones that formed a wall of three courses, in all about two feet thick, eight feet tall, and about sixteen feet of its length, as measured from the north end, was covered with the letters, arranged in wavy, nearly parallel and diagonal lines. The wall was traced and examined in many places for a distance of nearly a thousand feet, its course marked on the surface by stones like No. 1, projecting a few inches above the surface of the ground, and twenty-five or thirty feet apart. Seventy-five feet of the south end of the wall was bent at an angle of 15° to 20° east. The wall ended in a hollow of the hill.
>
> In March, 1891, the *Cleveland Express* printed a short account of the discovery, written by Mr. Carson of that place, who had seen the wall. In the *Sunday Sun*, New York, June 7th, I published a short notice of the find, with engravings made from my sketches made at the place, May 21st. . . .

The stone is dark-red sandstone, and the wall lies along the crest of a ridge of that kind of stone which trends north and south, flanked by limestone east and west, and extending from the Hiawassee River north to Chattanooga, south where it dips below the bed of the Tennessee River.

The surface of the west side of the inner course of stones is cut into rounded ridges with hollows between, and the characters are raised on the crest of the ridges, and are from two inches to three inches in width, with a few larger groups. Mr. J. Hampden Porter says, in a letter from Chatata, October 21st: "It is not a wall but a red sandstone ridge, faced with red, slaty, and yellow clays to an unknown depth. No implements and no traces of previous excavations have been found." The faces of the other course of stones are level and not cut into grooves. Between the courses is found a dark-red cement, which is probably formed of red clay with salts carried down by water.

Mr. Porter says: "As a rule inscriptions are intended to be real. . . . I do not remember any instance of a designed concealment like this."

The architect of the Pharos of Alexandria, Egypt, cut his name on the stone, covered it with plaster, and moulded Pharaoh's name in the covering. Time tore off the plaster and exposed the builder's name. This concealment in Tennessee may have been effected in a time of invasion or some great social calamity. Eight hundred and seventy-two characters have been examined, many of them duplicates, and a few imitations of animal forms, the moon and other objects. Accidental imitation of oriental alphabets are numerous.

The rock was chiseled in the form of letter intended, a hard cement worked in and raised above the surface, and a cement placed over the whole, against which the outer course of stones was placed, fitting closely. A piece of this covering cement with the letter-form in its surface is engraved here. The bird or other animal is the largest of that kind of figures that is found on the wall. Some of these forms recall those on the Dighton Rock, and may belong to the same age. How many other hidden inscriptions there may be in this, the geologically oldest continent, it is impossible to say, but delightful to conjecture. . . .

It is indeed interesting, if not delightful, to conjecture how many other inscriptions there may be on this continent. While there is no precise way of dating the mysteriously inscribed wall of Mr. Hooper, when one locates an object in a coal bed, he knows that he must be speaking of a passage of time in the neighborhood of 300 million years.

On April 2, 1897, a very peculiar piece of rock was removed from the Lehigh coal mine in Webster City, Iowa. The slab was found just under the sandstone, which was 130 feet beneath the surface.

The tablet was about two feet long by one foot wide and four inches thick. The surface was artistically carved in diamond-shaped squares, with the face of an old man in each square. Of the faces, all but two are looking toward the right. The features of each of the portraits were identical, with each bearing a strange mark in the shape of a dent in the forehead.

We can better understand why we as laymen seldom hear about such remarkable 300-million-year-old artifacts when we consider the difficulty that Dr. John C. Kraft, chairman of the department of geology at the University of Delaware, has undergone in his attempt to gain acceptance for a pendant that he believes to have been made on the North American continent anywhere from 12,000 to 10,000 years ago.

The five-and-one-half-inch-long piece of whelk shell bears the distinct carving of a woolly mammoth. It was dug out of a peat bog in Holly Oak, a town north of Wilmington, Delaware in 1864, and the majority of archaeologists have always considered the artifact to have been faked.

Dr. Kraft has taken cores of sediment from the bog area where the object was found (two highways and a railroad cross the site today), and two modern dating techniques reveal dates of 80,000 to 100,000 years before the present. Dr. Kraft maintains that there was a land surface existing in the bog area from 12,000 to 10,000 years ago, and he believes that the object, which he refers to as a pendant, could have come from either time period.

Although it is well accepted that mammoths roamed the United States until about 8,000 years ago, no association between man and the tusked giants is known in the northeastern United States at that time. Mammoth bones have been unearthed in Pennsylvania, New Jersey, and New York, but not in Delaware.

Dr. Kraft told *Science Digest* of June, 1976, that he believed much of the disbelief surrounding the pendant to have issued from the fact that the man who discovered the artifact had possessed such an abrasive personality. However, that same man, Dr. Hilborne T. Creeson, who had taught at Yale, was the finder of over 1,000 other Indian artifacts in addition to the controversial pendant, none of which were questioned as to their authenticity. The resistance toward accepting the artifact as genuine seems to lie in a stubborn "establishment" attitude toward recognizing an antiquity for sentient man on the North American continent greater than 30,000 years ago.

The carved shell now rests in the Smithsonian Institution in Washington, D.C. Dr. Kraft would very much like to do a radiocarbon dating of the artifact in an attempt to stifle some of the opposition, but it would require a hundred grams to do such a test.

"And that's almost the entire shell," Dr. Kraft explained.

"The Smithsonian's point is that it might be a fake, but if we destroy it, what have we done?"

It seems a point of some interest to ponder why certain scholars may find it an acceptable premise that primitive man may well have gathered around a mired mammoth and hurled stones at its skull to hasten its death and guarantee meat in the pot, but reject almost totally a creative, as well as a survival, impulse in those same humans. Science is willing to grant a desire to make soup of a bone, but denies the desire that may have inspired one of their number to carve a representation of their prey on a piece of shell or bone. It may not have been "art" as we understand it, after all, but a kind of magic that would either bestow upon the wearer of the pendant the strength of the mammoth or the ability to track it and kill it.

Another area of man's continuing efforts to control his environment, which the great number of men and women living today consider very new and strictly a province of our modern era, is that of medicine. To imagine the medicine of 3,000 years ago is to conjure up scenes of beating drums, singing chants, and devising torturous attempts to drive out evil spirits from infected teeth, brain tumors, and gangrenous limbs. Recent discoveries, however, provide us with quite a different picture of ancient medicine.

In 1972, Professor Andronik Jagharian, head of operative surgery at Yerevan Medical Institute in Armenia, stated that in his role as anthropologist he had examined skulls discovered in a lake bed near Yerevan and found that two of them bore evidence of intricate, delicate head surgery. Further investigation of the skulls revealed that they were at least 3,500 years old.

The skulls were first discovered when the erection of dams around Lake Sevan caused the water to recede and to disclose the ruins of an ancient city. Scientists identified the city as that of Ishtikunny, a settlement of folk called the Khurits.

According to Professor Jagharian, the first skull he examined was that of a 35-year-old woman. At some earlier age, she had evidently injured her head in such a manner that an opening was made in her skull. Although the wound did not appear to have damaged her brain, such a hole would certainly have involved the loss of a great deal of blood and would have exposed the brain to the possibility of easy injury.

"So those surgeons of 3,500 years ago made a plug from animal bone and inserted it into the hole," Professor Jagharian told journalists William Dick and Henry Gris. "The delicate operation was completely successful. I can tell from examining the skull that she lived for several years after, because I can see that her own bone had time to grow around the plug."

The second skull, also that of a woman, which Professor Jagharian examined bore evidence of an even more complex and delicate surgery. The anthropologist-surgeon guessed that a blunt object had somehow been forced into her skull.

"This kind of injury is extremely difficult to repair because the skull is made up of three layers of bone," Professor Jagharian explained to the journalists. "A sharp blow to the head produces splintering of the inner layers that is wider in diameter than the damage to the outer layer. Thus, to remove the splinters, it is necessary to cut a wider hole in the outer skull to get at the more extensive inner damage."

Professor Jagharian commented that even today such surgery would be considered a delicate and risky operation, but "those remarkable surgeons of 3,500 years ago performed it successfully. I can tell from bone growth that the woman lived about 15 years after the operation."

Russian scientists have found evidence that the surgeons of that dim and remote time performed operations with all the skill of modern medicine, including the use of anesthesia. It would appear that the scent of certain flowers was used to put people to sleep just as effectively as a general anesthetic is used today.

Professor Jagharian said that they had found the remnants of nearly 50 flowers and herbs that the ancients used for the purpose of general anesthesia.

"Considering the stone tools the ancient doctors had to work with," the scientist told Dick and Gris, "I would say they were technically superior to modern-day surgeons. For surgical tools, those early physicians used chisels made out of obsidian, a black stone that can be honed to great sharpness. We've found 4,000-year-old obsidian razors at Lake Sevan that are so sharp that they can still be used today."

Don Crabtree, of Kimberly, Idaho, is so confident of his

knowledge of the construction of surgical tools of the ancient Amerindian physicians that he permitted a local doctor to perform major surgery on him in October of 1975 with handmade obsidian tools.

"The ultimate experiment," as Crabtree termed it, involved the use of the obsidian knives to make an incision three-fourths the way around the trunk of his body to have a tumor removed from his lung. Crabtree, who has for the past 25 years been recreating the tools and implements of primitive man to the point where he is recognized as one of the world's top authorities on lithic (stone) technology, states that obsidian volcanic glass is a thousand times sharper than the platinum-plus blades used in other surgical implements. In Crabtree's expert opinion, the technique for making such tools is at least 10,000 years old.

The cutting surface of obsidian is so sharp that it does not bruise the cells, Crabtree states. Healing is more rapid and scarring is diminished. Crabtree is convinced that obsidian surfaces will eventually revolutionize surgery and could be especially valuable in plastic and cosmetic surgery.

Obsidian instruments were used by ancient Mayan surgeons for delivering caesarian births. According to Crabtree, the royal women of that culture were not permitted to give birth naturally.

For three or four hundred years the technique for making the volcanic glass tools was lost, but now Crabtree has reclaimed it for the modern world, and he is hopeful of encouraging more surgeons to experiment with obsidian knives.

The Egyptians used a contraceptive jelly that was applied on a wad of fibers and inserted deep into the vagina. The jelly was a mixture of honey, dates, and acacia spikes, which had been ground together in very fine consistency. The Western world was not to learn of this method of birth control until several thousand years later, and it was only in very recent years that it was learned that acacia spikes contain a gum that is deadly to sperm. When the gum is dissolved in a fluid, its active constituent is lactic acid, a familiar ingredient in many modern contraceptive jellies.

It was not until 1926 that modern science rediscovered the urine pregnancy test. In Ancient Egypt, women could take a pregnancy test at the earliest possible stages and at the same time determine the sex of the unborn child. The Egyptian laboratory techniques would take a sample of a woman's urine and soak bags of wheat and barley with the specimen. According to their observations, if the child was to be a boy, the growth of the wheat would be accelerated. The barley would be stimulated if a girl was to be born. In 1933, our own scientists confirmed by laboratory tests the acceleration of wheat and barley.

Even in the warm climes there are those who cannot bear continued exposure to the direct rays of the sun. Those Egyptians who followed the desert caravan routes across the Sahara found that they would gain extra protection from the sun by chewing a root called *ami-majos*. Modern research has shown that the root reinforced their skin pigmentation because it contains the active organic chemical compound referred to as 8-methoxypsorate.

In a compilation of Indian medical knowledge gathered by the first century A.D. court physician Charaka there is a pharmacopoeia that lists more than 500 herbal drugs. Included among the number is *Rauwolfia serpentina*, which derives its Latin name from the sixteenth-century German physician and botanist Leonhard Rauwolff, who identified the plant as a sedative.

In his *Lost Discoveries*, Colin Ronan states that the Indian physicians used the plant 1,500 years earlier "for colic, headaches, and, above all, as an anti-depressant—the 'medicine of sad men' it was called." Chemical analysis in modern laboratories has revealed that the plant contains "a number of pow-

erful alkaloids, including reserpine, a tranquilizing drug intro-
duced into Western psychiatric medicine only in the 1950s.''

Ronan goes on to point out that the ancient Indians also
practiced plastic surgery and utilized sutures (''stitches'') for
binding the edges of surgical wounds.

The *Susruta Samhita*, a medical book collected in the fifth
century A.D., describes how to use skin from the cheek or
forehead to replace a nose lost through accident or disease.
The same compilation also told how to sew surgical wounds
with curved needles made of bronze or bone. Such needles
were not adopted in the West until the nineteenth century.

Ronan tells of one ingenious method of suturing that the
Western world has never followed up—using large black Ben-
gali ants for joining intestinal wounds: ''The ants were placed
side by side and would clamp the edges of a wound in their
jaws; they were then decapitated and their bodies removed,
the heads remaining behind to dissolve away by the time the
wound had healed. The intestines, complete with their ma-
cabre suture, were then replaced and abdomen sewn up.''

As early as the sixth century B.C., the physician Susruta
was performing cataract removals. In the *Susruta Samhita*,
the Indian doctor, whom historians have named the ''Hippoc-
rates of India,'' after the Greek ''father of medicine,'' presents the precise details for cataract removal in what appears
to have been a routine operation (even though it took Western
practitioners another 2,000 years to dare attempt similar
procedures):

The patient—who has been fed and washed and carefully
tied—is placed on the ground.
Let the physician first warm the patient's eye with the breath
of his mouth. Then, rubbing gently with his thumb, he will de-
tect the impurity which has formed in the pupil.
The physician orders the patient to look down at his nose.
Holding the patient's head firmly, the physician takes a nee-

dle between his forefinger, middle finger and thumb. He carefully introduces it into the diseased eye—toward the pupil, on the side. He gently moves the needle back and forth and upward, pressing carefully against the eye of the patient. If he has probed correctly, there will be a sound and a drop of fluid will ooze from the eye, painlessly. Let the physician then moisten the eye with fresh mother's milk. Scratching the pupil with the tip of his needle, he must gradually press the impurity toward the nose. . . .

In the third century A.D. two remarkably accomplished Arab surgeons, who were renowned medical miracle workers, performed a leg transplant. According to official records of the Roman Catholic church, Cosmas and Damian, Christian converts and accomplished physicians famed for their cures, successfully removed the leg of a Roman nobleman and replaced it with a healthy leg taken from a black slave.

Journalist Ron Caylor, investigating the authenticity of the account, which is recorded in the *Enciclopedia Cristiana*, the official Italian Roman Catholic reference work, and is depicted on an ancient wood carving at the Cathedral of Palencia, Spain, quoted Dr. Jose Rivas Torres, a medical professor at Spain's Malaga University, as saying, "The evidence of the carving is perfectly clear to any surgeon. What it depicts is obviously a leg transplant, and the cathedral's records back this up. Modern medicine has not yet conquered the problem of rejection of foreign tissue by the human body that would make such a limb transplant possible. This is historical evidence that medicine was fantastically advanced centuries ago."

The two surgeons became Christian martyrs when Emperor Diocletian had them beheaded in 303 A.D. Monsignor Giovanni Ottieri of the Vatican Library in Rome stated that Vatican records prove that "Cosmas and Damian were sainted shortly after their deaths because of their medical miracles and their martyrdom."

Nor did ancient man neglect his teeth.

On January 23, 1970, Dr. Lucile E. St. Homy, of the Smithsonian Institution, and Dr. Richard T. Koritzer, a Glen Burnie, Maryland, dentist announced that they had recently discovered two "beautifully filled teeth" in a 1,000-year-old skull dug up 32 years ago near St. Louis, Missouri. According to the anthropologist and the dentist who made the dental discovery, the cementlike fillings constituted the "first evidence of a tooth preparation for therapeutic reasons in any prehistoric or ancient population."

We may one day soon learn that man has been involved in a cyclical pattern of relearning a great deal of what we moderns classify as advanced thinking and the finest of medical and dental care.

CHAPTER TEN

Floods, Fires, Catastrophes:

Armageddon Revisited

Each of the Amerindian tribes with which I am familiar cherishes legends that tell of their people emerging from the destruction that had been visited upon a former civilization. The majority of the accounts deal with the surviving peoples having escaped from a terrible flood, which immediately suggests both the biblical story of The Deluge and the Atlantis mythos.

The Yuchi Indians, who lived in what is now South Carolina and Georgia, tell of a big flood that drowned all but those who had been warned of the impending disaster. After the flood, goes the legend, the survivors attempted to build a high tower in which to take refuge should such a deluge again strike Earth. But such an enterprise brought only a differing of speech among the workers and a subsequent scattering of the people. (The latter account is highly reminiscent of the biblical Tower of Babel.)

The Arkansas Indians told of a destructive flood that was sent to Earth from God because of man's great wickedness.

The more pious survivors made their way to the North American continent in order to remain separate from others who might again become corrupt.

The Navajos say that their ancestors escaped the terrible flood through a long, hollow reed. When they reached safety, they were taken in spirit through space to visit other worlds, the Moon, the stars. After the waters had subsided on Earth, they returned to live on the mountains and in the cliffs. The world will again be destroyed if man cannot control his wickedness.

The Mandan Indians believed their ancestors rode out the flood in a big canoe, which came to rest on a mountain when the waters subsided.

The Delaware, or the Lenni-Lenapi, recounted the story of a continuing struggle between the men of Earth and the Snake-People. Resolving to destroy mankind, the Snake-People brought about a great rushing of water to drown all men. A female spirit helped some men to a boat and saw them to safety. The immigrants landed first in a cold country, but gradually worked their way to a more temperate land. In the meantime, the Snake-People migrated to the east and conquered a prosperous nation. Some of the Delaware remained in the new land, while others made their way back home.

The Popul Vuh, sacred book of the Central American Indians, states the flood myth in this way:

"Then the waters were agitated by the will of the Heart of Heaven, and a great inundation came . . . upon heads of these creatures. . . . They were engulfed, and a resinous thickness descended from heaven . . . the face of the Earth was obscured, and a heavy darkening rain commenced—rain by day and rain by night. There was heard a great noise above their heads, as if produced by fire. Then were men seen running, pushing each other, filled with despair. They wished to climb upon the trees, and the trees shook them off. They wished to enter into the caves, and the caves closed before them. Water and fire contributed to the universal ruin at the time of the last great cataclysm, which preceded the fourth creation.''

The Hopi legends also tell of the four worlds which were created and destroyed because of man's propensity to fall into errors of judgment and ways of moral corruption.

The principal point of each of the Amerindian myths of destruction and rebirth is that civilization is cyclical, continually being born, struggling toward a Golden Age, then slipping backward into moral morass, forward into its death throes . . . only to be reborn so that the process may begin once more.

The Seven Worlds legend has been revealed to few outside of those who are a part of the oral tradition of the Seneca. In my book *Medicine Talk* I included the legend in its entirety, which was translated by Twylah Nitsch, Repositor of Seneca Wisdom. For our purposes here, I will quote portions of sections after *Swen-i-o* (the Great Mystery) has created the heavens, Earth, and all creatures upon our terrestrial ball:

The First World

The nations of the First World emerged at the place where the Sun raised its head above the rim of the sky. At this place Mother Earth shared her gifts in great profusion. But the people at that time were not grateful for these gifts and caused a disease of waste to visit Nature Land.

Swen-i-o looked at man and arranged a time for the first decree: "You are creatures of nature, created by me to live always in true harmony. Wisdom, if learned, is balanced of life. Breaking this law, breeds misery and strife. The Great Spirit has spoken."

For a time, the people were impressed by the Great Revelation they had heard, but they soon found it very hard to follow the decree of the Great Spirit.

As time passed, the decree was forgotten, and *Swen-i-o* arranged for a cleansing of the First World.

He placed a blanket of protection over those creatures who honored his decree. He ordered the Sun to use its power in cleansing the First World. The power of the Sun caused the devastation of the First World.

The Dawn of the Second World

The lessons learned from the acts of the people who perished in the First World remained in the minds of those who were saved. Carefully they populated the Second World. . . . Their culture was superb, and it spread rapidly throughout the Second World.

Migrations moved toward the North, a place of total whiteness; to the South, a place of total darkness; with the nations adapting to the environment of these places. Their outer skin became faded where it was cold, and dark where it was hot. Migrations followed the Sun as he travelled the path of the Sky Dome from East to West.

Before long, it became evident that the people of the Second World were following in the footsteps of their predecessors, who had inhabited the First World. The wanton waste of the gifts of Mother Earth and the careless imbalance of their lives brought on misery and strife that gripped the world in a disease of destruction. Those who still honored the decree of the Great Spirit were given a blanket of protection, and the cleansing of the Second World was begun.

Swen-i-o ordered the Sun to withdraw its warmth from the face of Mother Earth, leaving only the Moon to exert his power upon Nature Land. The lesser light of the Moon was unable to warm Mother Earth. A state of cold settled upon Nature Land. This caused the devastation of the Second World.

The Third World

The Third World was inhabited by people and creatures with gifts and abilities that surpassed the two previous worlds. They spread their influence along the path of the Sun, establishing magnificent civilizations and cultures, populating more than half of the world. Four races had evolved as a result of migrations: the white, the red, the yellow, and the black—their complexions and physical characteristics having adapted to the environment in which they lived.

During the Third World the four races became more aware of the laws that governed Nature Land, and they made some effort to learn about its mystery. For this reason their civilization flourished for a longer period than the first and second worlds. But in spite of their knowledge, they became forgetful, and they consistently brought disruption upon the gifts of Mother Earth.

For the third time, those who honored the decree of *Swen-i-o* were placed under the blanket of protection. Water, the third creation of *Swen-i-o*, was responsible for the cleansing of the Third World.

In the Mind of Swen-i-o Came the Dawn of the Fourth World

The migrations of the Fourth World completed the population of the universe from East to West. The greatest span of existence was experienced by the people evolving in the Fourth World, because this world was the Middle World. Those whose evolvement had reached the awakening period were willing to share their knowledge with others. They began to keep records; but the greatest records were still in the minds of the generations who had lived under the Blanket of Protection and who still honored the decree of the Great Spirit. These people had evolved along the thread that connected them to *Swen-i-o*, the creator.

During the Fourth World, the inhabitants became aware of the universal stream that revolved around the world, and they learned the wisdom of enlightenment. Through the minds of these people the Secret of the Ages was recorded.

Unfortunately, too many still pursued the materialistic path, spreading misery and doom among the inhabitants. . . .

It became evident that the Fourth World would have to undergo a cleansing period to renew the gifts to Mother Earth, just as the three previous worlds had. The cleansing of the Fourth World was exerted by the combined efforts of the Sun, the Moon, and Water upon Mother Earth. The Fourth World's corruption had been the greatest—therefore, its need for renewal, the greatest.

In the Mind of Swen-i-o was the Dawn of the Fifth World

The greatest strides in understanding took place in the Fifth World. The era of the awakening had become established, and man found self-satisfaction in sharing his gifts and abilities with others. The Records of the Ages were beging uncovered, and the false documents of man were being corrected.

The duration of the Fifth World was short compared to the previous worlds. Man had passed through many environmental experiences. His lessons were extremely difficult, but he had achieved self-mastery. . . . However, he was not yet convinced of the duality of his nature regarding the function of his spiritual mind over the physical body. Wars within his attitudes and thoughts still festered in his mind, creating injustices upon the gifts of Mother Earth.

Man was having difficulty practicing the decree of *Swen-i-o* that had been revealed in the First World. Repeatedly, there had been messengers of the Great Spirit to remind the people of the wisdom of harmony. Yet they were unable to perpetuate their beliefs after the general disturbances of their way of life in the latter days of the Fourth World.

Mother Earth was again in need of a renewal of her gifts. The cleansing was exerted by the powers of the Sun and Moon and was completed in the mind of *Swen-i-o* the Creator.

The Dawn of the Sixth World

The Sixth World had the shortest evolvement period. It was the world that opened the eyes of man. . . . He . . . recognized the necessity of fitting into a pattern that functioned in unison with his world. As yet, he had not fully accepted the laws of nature as his guide. His life at times was still governed by his own selfish thinking. There had to be one more cleansing to renew the gifts of Mother Earth, before man truly understood and could practice his purpose in life.

> For all six worlds, man had wreaked havoc upon
> himself, his fellow man, and the creatures
> of Nature.
> Now he stood at the threshold of perfection,
> awaiting the wisdom of the ages to
> penetrate his mind.

The cleansing of the Sixth World was exerted by the power of the Moon, followed by the heating properties of the Water, which paved the way for the dawn of the Seventh World.

The Seventh World

> In the Seventh World, the Happy Hunting
> Ground,
> Man saw beauty everywhere.
> He listened to the music of the Universe.
> And sang his part in the chorus.
> He felt love for *Swen-i-o* and for his fellow-man.
> He shared his gifts and abilities with others.
> He made the Seventh World a place of peace and
> happiness.

The final cleansing had been completed, and man's life was guided by a spiritual light, the same light that is in the mind of *Swen-i-o*, the Great Spirit.

The Seneca Legend of the Seven Worlds says that the world of man has relived the traumatic experiences of birth, death, and rebirth six times before and that we stand on the brink of destruction prior to entering the final world in our evolutionary cycle.

The Hopi agree and say that we are about to enter the final world after a last great war, a war that will be ". . . a spiritual conflict with material matters. Material matters will be destroyed by spiritual beings who will remain to create one world and one nation under one power, that of the Creator."

The Armageddon of spiritual against material will occur when the "Saquasohuh (Blue Star) Kachina"—now represented by a far away, and yet visible, blue star—makes its appearance. The process of emergence into the final world, however, has already begun, the Hopis tell us.

For the Amerindian traditionalist, the destructions of the previous worlds have all been a necessary part of mankind's spiritual evolution. Man forgets the lessons of the Great Spirit and falls away to rely upon his own feeble devices. When this state of affairs comes to pass, the Great Spirit causes a time of Great Purification, which cleanses the Earth Mother for a new epoch, a new world.

It is obvious that the Amerindian view of the "worlds before our own" is one of catastrophism, that there have been a succession of cultures created, leveled, then recreated. As Immanuel Velikovsky has pointed out, the vast majority of ancient texts and the oral traditions of most primitive peoples "deal specifically with the phenomenon of catastrophism."

It is really too simplistic to state that all those scientists who embrace tenets of the catastrophist school believe that the world is only 6,004 years old and that all uniformitarian geologists believe our planet to be four-and-one-half billion years old. In the February, 1975 issue of *Natural History*, Stephen Jay Gould argues that modern geology is really a blend of concepts from both the uniformitarians and the catastrophists.

In his article, "Catastrophes and Steady State Earth," Gould writes:

"A 6,000-year-old earth does require a belief in catastrophes to compress the geologic record into so short a time. But the converse is decidedly not true: a belief in catastrophes does not dictate a 6,000-year-old earth. The earth may be 4.5 or, for that matter, 100 billion years old and still build its mountains with great rapidity."

Gould admits: "The geologic record does seem to record catastrophes: rocks are fractured and contorted' whole faunas are wiped out . . . ."

The classic uniformitarian position, as developed by Charles Lyell in 1830 in the first volume of his revolutionary *Principles of Geology,* are summarized by Gould in the following manner:

1. Natural laws are constant (uniform in space and time). . . . This is not a statement about the world; it is an *a priori* claim of method that scientists must make in order to proceed with any analysis of the past. . . .
2. Processes now operating to mold the earth's surface should be invoked to explain the events of the past (uniformity of process through time). . . . This again is not an argument about the world; it is a statement about scientific procedure. . . .
3. Geologic change is slow, dual, and steady, not cataclysmic or paroxysmal. . . .
4. The earth has been fundamentally the same since its formation (uniformity of material conditions). . . .

Lyell's vision of a uniformitarian Earth caused him to reconcile the appearance of direction with dynamic constancy in the history of life by proposing that the entire fossil record represents but one part of a "great year"—a grand cycle that will occur again when ". . . the huge iguanodon might reappear in the woods, and the ichthyosaur in the sea, while the pterodactyle might flit again through umbrageous groves of tree ferns."

The catastrophists pursued the literal view. "They saw direction in the history of life, and they believed it," Gould remarks. "In retrospect, they were right.

"Modern geology is really an even mixture of two scientific schools—Lyell's original, rigid uniformitarianism and the scientific catastrophism of Cuvier and Agassiz. We accept Lyell's first two uniformities, but so did the catastrophists. Lyell's third uniformity appropriately derigidified is his great substantive contribution; his fourth (and most important) uniformity has been graciously forgotten. . . ."

Re-examining the Seneca legend of the Seven Worlds from a catastrophist's point of view, we find that the First World was destroyed by "the power of the Sun," fire, perhaps, or, might we say, radioactivity?

The Second World, with only the Moon to warm the Earth, may be referring to a great Ice Age, either the one with which we assume familiarity or one in times long before our accepted geologic records.

The destruction of the Third World was brought about by water, perhaps caused by great shifts in the Earth's crust.

The Fourth World is said to have been one of great intellectual accomplishment. There were great migrations, a period of enlightenment and awakening, technological accomplishment. The very Secret of the Ages was recorded.

Could this have been the "world before our own" that bequeathed many of the legendary wonders of the ancient world that appear so out of context with our evaluations of the early stages of our epoch? The statement that the Fourth World was destroyed by the combined efforts of Sun, Moon, and Water—fire, ice, flood—could suggest the horrible devastation of nuclear power to the point where some nations were totally destroyed and even submerged.

The Fifth World may have been but a prolongation of the Fourth. Seneca legend records that although awakening of the spirit was evident and many false documents were corrected, the duration of the Fifth World was brief.

We may, in the Fourth World, be remembering a global war of terrible nuclear conflict. An uneasy peace is reached by the survivors, and men of good will on both sides attempt to establish a time of harmony and resurrection. But the sparks of enmity still smolder, and they are ignited by war hawks in a final paroxysm of destruction.

The Sixth World may have been the antediluvian world to which the Old Testament refers, a world once again destroyed by flood.

Or, according to some interpretations, it may be the world in which we are presently striving; for it does not seem as though we have yet attained the Seventh World wherein man "sees beauty everywhere," is attuned to "the music of the Universe," or feels love for God and our fellow man. And if the "power of the Moon" refers to a time of destruction of cold, many authorities have stated their conviction that we are entering a cold-dry cycle of history preparatory to another Ice Age.

Is there any substantiating evidence for the Seneca legends of a world having been thrice destroyed by the power of the Sun, i.e., fiery, nuclear explosions?

Albion W. Hart, one of the first engineers to graduate from Massachusetts Institute of Technology, was assigned an engineering project in the interior of Africa. While he and his men were traveling to an almost inaccessible region, they had first to across a great expanse of desert.

"At the time he was puzzled and quite unable to explain a large expanse of greenish glass which covered the sands as far as he could see," writes Margarethe Casson in No. 396, 1972, of *Rocks and Minerals.* "Later on, during his life . . . he passed by the White Sands area after the first atomic explosion there, and he recognized the same type of silica fusion which he had seen fifty years earlier in the African desert."

In his article "Strange Fire on the Earth," *Creation Research Society Quarterly,* December, 1975, Erich A. von Fange examines certain scriptural accounts and various ancient legends that have to do with great destruction by fire. It is von Fange's thesis that it is possible that "atomic chain reactions boiled on the earth in the distant past."

French researchers discovered the evidence of prehistoric spontaneous nuclear reaction at the Oklo mine, Pierrelatte, in Gabon, Africa. The scientists found that the ore of this mine contained abnormally low proportions of U_{235} such as are found only in depleted uranium fuel taken from atomic reactors. According to those who examined the mine, the ore also contained "four rare elements in forms similar to those found in depleted uranium."

Although the modern world did not experience atomic reaction until the 1940s, nuclear effects may have occurred in earlier times, leaving behind as physical evidence sand melted into glass in certain desert areas, hill forts with vitrified portions of stone walls, or the remains of ancient cities destroyed by what appears to have been intense heat beyond that of the torches of invading armies.

Those who have encountered this rather awe-inspiring aspect of prehistory stress the point that they are not referring to catastrophes that were caused by volcanoes, by lightning, by conflagrations set by man, or by crashing comets.

For example, there are ancient ruins in Arabia that date back to the time when the southern part of the peninsula was fertile and well-watered. In western Arabia, there are 28 fields of scorched and shattered stones that cover as many as 7,000 square miles each. The stones are sharp-edged, densely grouped, and burned black. They are not volcanic in origin, but appear to date from the period when Arabia was a lush and fruitful land that suddenly became scorched into an instant desert.

It was an intense blast of heat, coupled with other catastrophic events, which transformed a tropical region of heavy vegetation, abundant rainfall, and several large rivers into what we know today as the Sahara Desert. Scientists have discovered areas of the desert in which soils which once knew the cultivated influence of plow and farmer are now covered by a thin layer of sand. Researchers have also found an enormous reservoir of water below the parched desert area. The source of such a large deposit of water could only have been the heavy rains from the period of time before a fiery devastation consumed the lush vegetation of the area.

In 1947, in the Euphrates valley of southern Iraq, where

certain traditions maintain that human life began, exploratory digging unearthed a layer of fused, green glass before the cultural level of Sumer. Again, this several-thousand-year-old fused glass bore a resemblance to nothing more than the desert floor at White Sands, New Mexico, after the nuclear blasts melted the sand and rock.

The Red Chinese have conducted atomic tests, near Lob Nor Lake in the Gobi Desert, which have left large patches of the area covered with vitreous sand. But the Gobi has a number of other areas of glassy sand which have been known for thousands of years.

In the United States, the Mohave Desert has large, circular or polygonal areas that are coated with a hard substance very much like opaque glass.

There are ancient hill forts and towers in Scotland, Ireland, and England in which the stoneworks have become calcined because of great heat that has been applied. There is no way that lightning could have caused such effects.

"Hill forts of the west Atlantic coast fringe, from the Lofoten Islands off northern Norway to the Canary Islands off northwest Africa, became so-called fused forts, and the piled boulders of their circular walls have been turned to glass like frozen treacle on the outer portions of the wall facing the west," writes Erich A. von Fange in the *Creation Research Society Quarterly*. "Some intense heat created the same effect on the inner sides of the eastern part of the circle. Similar vitrifications have been reported from the Western Pacific.

"The same phenomenon has been observed in the mounds and barrows of the British Isles. . . . The stones of the innermost cell of a long barrow near Maughold on the Isle of Man have been fused together like the mysterious vitrified towers of Scotland and elsewhere. . . ."

One of the oldest cities in the world is thought to be Catal Huyuk in south-central Turkey. The city appeared, according to first known evidence, to have been fully civilized and then, suddenly, to have died out.

Archaeologists found thick layers of burned brick at one of the levels, called VIa. The blocks had been fused together by such intense heat that its effects penetrated to a depth more than a meter below the level of the floors where it carbonized the earth, the skeletal remains of the dead, and the burial gifts that had been interred with them. All bacterial decay had been halted by the tremendous heat.

When a large ziggurat in Babylonia was excavated, it gave the appearance of having been struck by a terrible fire that had split it down to its foundation. In other parts of the ruins, large sections of brickwork had been scorched into a vitrified state. Several masses of brickwork had been rendered into a completely molten state. Even large boulders found in the vicinity of the ruins had been vitrified.

The royal buildings at a north Syrian site known as Alalakh or Atchana had been burned so completely that the very core of the thick walls had bright red, crumbling mud-bricks. The mud and lime wall plaster had become vitrified and basalt wall slabs had, in some areas, actually melted.

Between India's Ganges River and the Rajmahal Hills are scorched ruins containing large masses of stone that have been fused and hollowed. Certain travelers who venture to the heart of the Indian forests have reported ruins of cities in which the walls have become huge slabs of crystal, due to some intense heat.

While exploring Death Valley in 1850, an adventurer, William Walker, claimed to have come upon the ruins of an ancient city. An end of a large building within the rubble had had its stones melted and vitrified.

Today, of course, the area is a sterile desert; but in an earlier chapter, we have already mentioned the evidence which indicated that Death Valley was once a tropical paradise, populated by an unknown race of giants.

Walker stated that the entire region between the Gila and St. John rivers was spotted with ruins. Each of the ancient settlements was burned out and vitrified in part by fire intense enough to have liquefied rock. Paving blocks and stone houses were split with huge cracks, as if seared by some monstrous sword of fire.

The site of *Sete Cidades* (Seven Cities), Brazil, was first charted in 1928. The ruins, located near the equator in the Province of Piaui, are, according to Erich A. von Fange, "a monstrous chaos as one might imagine Gomorrah was. The stones are dried out, destroyed, melted."

Although no excavation has yet been conducted at the site, there is no geological explanation to explain the lumps of melted rock that dot the surrounding plains.

If we expand the catastrophic theory of history to include the allegation that there have been a number of highly cultured and technological civilizations in prehistory which were almost completely destroyed by a great disaster of some kind, may we also offer the suggestion that at least some of those civilizations might have destroyed themselves in a nuclear holocaust of their tragic creation?

In the holy books and legends of many ancient peoples we find innumerable accounts of wars between the heavens and the earth. Cosmic revolutions and civil wars were said to rent and to split the prehistoric worlds on several occasions. More than one Sodom and Gomorrah exploded so that "the smoke rose up like that from a mighty furnace," and references to their destruction are found in the scriptures of Hindu and Hebrew and in the myths of people as diverse as the American Indian and the African.

Each of the legends has the extant civilization being decimated. Its governments are rendered impotent, its commerce abandoned, its cultural attributes forgotten, its cities crumbled to rubble. But always a remnant of its people survive. Enough human seed is retained to perpetuate the stubborn and striving species. The cycle of the death and rebirth of civilization is maintained. Man returns to the primitive to relearn the basics, to recall the essentials. At the same time that he is once again mastering the elementary lessons of survival, he is re-establishing an understanding of his physical body's niche in the web of life and his ethereal psyche's position in the one-ness of the holy, the divine, the cosmic. It is as if man must continue to replicate the progression of hunter to farmer to merchant to scientist-philosopher until he sets it right and graduates from ape to angel.

The traditions of the American Indian state that periodical-ly the planet, the Earth Mother, purges herself in a time of Great Purification. The Medicine People say that this cleans-ing occurs, not because man is evil and needs punishing, but according to a natural cycle.

The Medicine People state that another time of purification has rolled around on the great cosmic calendar. We are about to enter a time of earthquakes, vulcanism, and dramatic earth changes. We can do no more to prevent the cataclysms before us than we can prevent the advent of winter. But just as a knowledge of the past tells us that, regardless of our fears, winter will come, it also informs us that we can prepare for it. In like manner can we begin to make preparations for the Great Purification and to plan ahead for the planetary spring time that awaits the humanity that endures.

To this end, some devoted men and women are establishing various repositories of books, records, paintings, motion pic-tures, and other artifacts of our knowledge and art so that fu-ture generations might discover those things that were best in our society—and so those who survive the Earth Mother's fierce trembling and shaking might have some cultural tools with which to work.

Once again man's science has the nuclear power available to transform our cities into those tragically familiar patches of fused green glass.

Must we scorch ourselves to oblivion in the madness of a nuclear holocaust?

Must our culture become the subject of mystery and debate for our faceless ancestors?

Will a future civilization speculate about whether or not a

177

people of some degree of sophistication actually existed on the North American continent in an epoch which their time-perspective will term "prehistoric"?

I have come to believe that civilization on Earth has been cyclical, that there have been highly evolved human or hominid cultures beyond our present epoch, that within man's racial memory or collective unconscious may lie half-forgotten memories of these worlds before our own.

Such mythic kingdoms as Atlantis may have become but a symbol of those previous worlds, a metaphor of man's brief glory before the dust of an eternity that makes the most long-lived empires seem but an eye-blink of God.

We have recently heard a great deal from the more mystically oriented members of our society that we stand now at the dawning of the Age of Aquarius, a new age of harmony, peace, love, and understanding. Would it really be an affront to our Space Age technology and our sophisticated modern intellects to consider that before each new age is permitted to dawn, the preceding age must be set into obscurity?

There is something within man that seems to drive him to find law, order, structure, and pattern in his world, his universe. Man's science is founded on the repeatability of experiments. His religion is built on the promise of a Divine plan and order.

I believe man knows, on an intuitive level, that time is cyclical and unwinds in a spiral, but he persists in his attempts to structure a linear history—a time that proceeds in a straight line, devoid of revolving patterns and returning seasons. Such things as stock and commodity prices, industrial production, war, depressions, civil riots, and geologic catastrophes are considered blights that occur without structure, without demonstrable pattern.

Recently, our experts on energy, agriculture, population, and world economy have predicted bankruptcy, social breakdown, and starvation for as many as one billion people by the time the words in this book have been published. Some thirty nations, mostly in Africa, South Asia, and the Central American–Caribbean area, will be among the first to be affected in what will soon be global famine.

In the 1930s and '40s, Professor Raymond Wheeler, who was head of the psychology department at the University of Kansas, invested over twenty years' time and the efforts of a staff of over two hundred to compile incredibly detailed records of 3,000 years of weather and the cycles that run through them. Nearly two million separate pieces of information about weather-in-history were entered on cards and were supplemented with maps and charts.

Wheeler was able to chart types of governments, human achievements, wars, and shifts in cultural styles from one extreme to another and back again. He was able to isolate definite patterns of human behavior as men and women reacted to climate changes.

He discovered a 100-year cycle, which is divided into four almost equal parts, that demonstrates that man has behaved differently—but *predictably*—during periods of warm-wet, warm-dry, cold-wet, and cold-dry weather. Wars, depressions, revolutions, cataclysmic events—together with tastes in architecture, musical expression, poetic metre, and the length of hemlines—have occurred at evenly spaced intervals. Wheeler's monumental research provides us with demonstratable bases for predicting what will happen in the years that lie ahead.

Astronomer Selby Maxwell, science editor of the *Chicago Tribune* in the 1920s, discovered a weather-energy cycle which has proved to be the basic cycle that governs all weather—past, present, and future. Maxwell determined the correct time lags which cause the turbulent upper-air masses to act in a predetermined manner. Crucial to Maxwell's method for predicting the weather was his revelation that all cycles of the same length turn at the same time and that all cycles are related in one way or another to this basic energy cycle.

By recognizing that patterns of time do recur at rhythmic

intervals, man can chart potential danger periods. He can organize his business, labor, and governmental agencies to produce and to distribute goods according to the best periods for growth. He can plan ahead for droughts and famines and store foodstuffs for the lean years. He can attempt to stockpile goods for periods of depression and inflation.

All of man's magnificent accomplishments, all the fine products of his reason—together with his animalistic, insane behavior that results in wars and riots—may actually be due to energy waves, cyclic forces beyond man's control.

One level of man's consciousness may be attuned to something such as electromagnetic impulses, which even though he cannot hear them or define them, do dramatically affect him and cause many of the crises of existence to which he is subjected.

In the book that he wrote with Og Mandino (*Cycles: The Mysterious Forces that Trigger Events*), Edward R. Dewey states that a "basic secret of nature" is that man is surrounded by cyclic forces which bounce men and women like marionettes on a string.

"Since it is demeaning to his self-esteem," Dewey writes, "it is perfectly understandable that man should resist any hypothesis that holds that his life and his universe vibrate in rhythms that are regular and at least partially predictable and are caused by a force or forces still unknown and possibly uncontrollable by him.

"Nevertheless, the evidence that man is not one step down from the angels, sublimely in command of himself and his world, continues to accumulate. He is more like a character in a Punch and Judy show, pulled this way and that by environmental forces. And he will continue to be so manipulated until he solves the mystery of these forces. Only then will he be able to cut the strings and become himself."

Perhaps man cannot control Nature, as he has for so long smugly believed, but he may certainly adapt himself to Nature's whims and its cyclic forces.

The cyclic forces are inevitable, but by knowing about them in advance through such research as the Wheeler-Maxwell discoveries, we may exercise our freedom of choice and hold the cyclic results subject to our will. Knowledge of what is ahead for us on the recurring pattern of time can enable us to prepare for adversity.

"I have always insisted that the outlook for man is not fatalistic, integrated though he is with his environment and subject profoundly to climatic influences," Dr. Wheeler stated. "There is no excuse, whatsoever, for becoming an environmental determinist—at least of the kind known to the history of science."

Naturally, everything that happens in this world must have a reason. The science and philosophy of the past have given to humanity differing points of view regarding reasons for events.

"One world view supposes that the universe operates in accordance with blind, mechanical laws, that the forces of nature are physical (in the sense of material forces), based on matter as the ultimate reality.

"If this philosophy is carried out consistently to its natural conclusion, as certain philosophers and scientists of the past have done, there is no place for the mind in such a world; there is no place for human values; there is no place for God," Dr. Wheeler said. "The universe is a vast, meaningless machine, having no purpose, going nowhere. Reality in the end is but a fortuitous concourse of whirling atoms. Laws and principles mean nothing except for purposes of prediction. The human organism, like everything else in the world, is but a mere machine, a robot, whose feelings, thoughts, and longings have no more significance than the bumping of one atom into another."

A second means of apprehending one's environment is the organic and idealistic view, as opposed to the mechanical and materialistic one.

The organic view maintains that the laws of nature are not

181

wholly mechanistic. Neither man nor the universe is a machine.

The organic view is a positive one. The universe means just what the term implies: *uni*, "one," and *versum*, "combined or turned into." The universe is One, a Unity. This means that all its parts, from the humble atom to man himself, are interrelated in a common whole; everything in the universe obeys the same ultimate laws.

"Modern science has revived the ancient concept of teleology (the view that everything in the world has a purpose), but in clothes so different that no one except the philosopher would be likely to recognize it," Dr. Wheeler said. "According to modern sciences, no activity can transpire in the world, there can be no motion of any kind, unless nature as a whole obeys laws of equilibrium.

"Activities have *direction*, or they cannot occur. They are headed somewhere. There is activity only when there is an assumed difference in potential between two points in a unified system of energy. Then the activity proceeds from the position of the higher to the position of the lower potential until the difference between the two is resolved; that is, until the system reaches equilibrium.

"Thus, while activity is going on, the state of equilibrium is in the *future*, but the events that are now occurring are doing so with reference to that future state.

"In other words, the conditioning factor—or at least one conditioning factor that is assumed to be necessary for the action—lies in the future of the events in question. The event is headed toward the goal of future equilibrium. The purpose of the event is the restoration of the system to equilibrium."

Everything in the universe has to follow the same laws, or the universe would lack unity; there would be no integration; there would be no universe.

Human ideals are objectives, perfections of one kind or another, that, by definition, lie ahead of human action. If these ideals could be reached they would not be ideals. Like the condition of perfect equilibrium of all the physical forces in the universe, they are forever in the future—but that future is forever giving direction to what is happening now.

"It is neither elevating the atom nor degrading man to assume that every object in the universe obeys the same laws," Dr. Wheeler assures us, "for the status of each object is defined, not by the laws it obeys, but by its function, or purpose, in a system where every part obeys the same laws of the whole. The functions or purposes of the two may be vastly different under the same laws.

"The human being has intelligence, which the atom or the rain or the heat or the cold does not have. Intelligence has a place, a purpose, in the world. What better purpose could it have than that of trying to ascertain the meaning of life and of the universe as the abode of man?"

Human qualities, then, are not passive victims of environmental forces. Human behavior is a force acting *against* environmental pressures, just as much as environmental pressures are forces acting upon man.

Man may eventually be able to create enough of an artificial climate in which to live to keep him from going to extremes under the influence of natural climate. By air conditioning, for example, he may be able to reduce in considerable measure the lethargy on the one hand, and the fanaticism on the other, which have been associated with hot drought epochs. He may be able to minimize the heretofore uncontrolled belligerence that has been associated with climatic transitions. He may become enlightened enough to eliminate the civil wars of colder times.

By submitting to the gas laws, man built engines that were capable of carrying him around the world on land, sea, and air.

By submitting to the laws of electricity, man gave himself light and heat, and instruments of communication, which could bring the people of different continents together as if there were no time or space.

By submitting to the inevitable facts of health and disease, man has reduced infant mortality, prolonged his life, conquered devastating plagues, enabled the lame to walk, the deaf to hear, and the blind to see.

Why then should man not control his destiny on a larger scale?

Wars will rage; whole civilizations will continue to collapse in internal misery and bloodshed until the natural causes of these events are known and recognized.

With an optimism predicated on a careful assessment of his research data, Dr. Wheeler stated, "There is emancipation ahead through a control of human nature, even as there was emancipation ahead when it became possible to control forces in the physical and biological world."

Even now as I write these final pages, the already abundant omens of transition and transformation are increasing. And Old World is dying all around us. A New World is being born. Another period of cleansing seems about to begin.

If there are to be any worlds *after* our own, we must commit ourselves to knowledge, understanding, and love so that we may survive the Armageddon that may be necessary to cleanse the Earth for a new epoch, a new age, a new turn of the wheel of civilization.

If a time of purgation is a predetermined aspect of a natural cycle, let us pray that our science will choose to utilize nuclear power as a means of survival against the approaching cataclysms rather than a means of total species annihilation. Nuclear energy need be no more a destructive demon than is electricity.

With the proper safeguards, our cities, our vehicles of transportation, our homes, and our day-to-day existences can be both powered and made better by intelligently applied nuclear technology. Rather than a radioactive Fifth Horseman of the Apocalypse, nuclear energy might be a powerful angel that can maintain our species through the time of transition we shall encounter in the next two decades. If we truly apply wisdom, a fused and united humanity, rather than a layer of fused green glass, might become the symbol of our epoch most remembered by the unborn inheritors of our planet. If we can at last learn to live in love, we shall bequeath a legacy of peace, rather than an enigma of our origins, to those men and women who will people the worlds after our own.

Benton, Jesse James. *Cow by the Tail.* Boston, Houghton Mifflin Co., 1943.

Burdick, Clifford. *Footprints in the Sands of Time.* Bible-Science Association, Box 1016, Caldwell, Idaho 83605, undated, circa 1975.

Cousins, Frank W. *Fossil Man.* New York, Praeger, 1966. Evolution Protest Movement (rev. ed.), 1971.

Däniken, Erich von. *Chariots of the Gods?* New York, Bantam, 1969.

Dewey, Edward R. with Mandino, Og. *Cycles: The Mysterious Forces That Trigger Events.* New York, Hawthorn Books, 1971.

Donnelly, Ignatius. *Atlantis: The Antediluvian World.* 1882.

Josyer, G. R. *Vimankia Shastra.* Translation of *Maharishi Bharadwaja,* Coronation Press of Mysore, 1973.

Lyell, Charles, Sir. *Principles of Geology.* London, John Murray, 1830–33 (3 v.).

Lyman, Robert Ray, Sr. *Forbidden Land, 1614–1895.* Coudersport, Pa., Potter Enterprise, 1971.

Miller, Francis Trevelyan. *The World in the Air.* New York, Putnam, 1930.

Neuburger, Albert. *Technical Arts and Sciences of the Ancients.* New York, Macmillan, 1930.

Renfrew, Colin. *Before Civilization.* New York, Knopf, 1973.

Ronan, Colin A. *Lost Discoveries.* New York, McGraw-Hill Book Co., 1973.

Steiger, Brad. *Medicine Talk.* Garden City, N.Y., Doubleday, 1975.

Stutzer, Otto. *Geology of Coal.* Chicago, University of Chicago Press, 1940.

Thompson, John Eric Sidney. *Rise and Fall of Maya Civilization.* Norman, University of Oklahoma Press, 1954.

Tompkins, Peter. *Secrets of the Great Pyramid.* New York, Harper & Row, 1971.

BIBLIOGRAPHY

COMPILED BY RONALD CALAIS

Creation/Evolution

Bernhard, R. "Thinking the Unthinkable—Are Evolutionists Wrong?" *Science Creation Research*, 4, 1969.
———. "Heresy In The Halls Of Biology." *Science Creation Research*, Nov. 1964.
Bethel, Tom. "Darwin's Mistake." *Harper's Magazine*, Feb. 1976.
Ciparick, J. "Myths & Models: Problems of Origin." *The Science Teacher*, Jan. 1973.
Clark, R. D. "The Origin of Life." *Journal of the Victoria Institute*, Winter, 1968.
Cloud, P. " 'Science Creationism': A New Inquisition?" *The Humanist*, Jan. 1977.
Eiseley, L. "Was Darwin Wrong about the Human Brain?" *Harper's Magazine*, Nov. 1955.
Gish, Duane, *Speculations and Experiments on the Origin of Life*. (Technical Monograph), No. 1, 1972.

————. "Creation, Evolution and Historical Evidence." *American Biology Teacher,* March, 1973.

————. *Evolution—the Fossils Say No!* San Diego, 1972.

Goldschmidt, R. B. "Evolution as Viewed By One Geneticist." *American Scientist,* 40, 1952.

Gould, S.J. "A Threat to Darwinism." *Natural History,* Dec. 1975.

————. "Reverend Burnet's Dirty Little Planet." *Natural History,* April 1975.

Grene, Marjorie. "The Faith of Darwinism." *Encounter,* Nov. 1959.

Johnson, W.G. "On the So-Called 'Science-Religion' Conflict." *Science Education,* 57, 1973.

Martin, C.P. "A Non-Geneticist Looks at Evolution." *American Scientist,* 41, 1953.

Medawar, Peter. *Mathematical Challenges to Neo-Darwinian Interpretation of Evolution.* Philadelphia, 1967.

Moore, John. "Evolution, Creation and the Scientific Method." *American Biology Teacher,* Jan. 1973.

Morris, Henry. *Scientific Creationism.* San Diego, 1974.

Mulfinger, G. *The Flood and The Fossils.* Greenville, N.C. 1974.

Nelkin, D. "The Science-Textbook Controversies." *Scientific American,* April 1976.

Orlich, D. et àl. "Creation in the Science Classroom." *The Science Teacher,* May 1975.

Patterson, R. "Life Not a Chemical Accident." *Science Digest,* 76, 1974.

Salisbury, Frank. "Doubts about the Modern Synthetic Evolution Theory." *American Biology Teacher,* Sept. 1971.

Shorey, Paul. "Evolution, a Conservative's Apology." *Atlantic Monthly,* Oct. 1928.

Simpson, George. "The World into which Darwin Led Us." *Science,* April 1, 1960.

Chronology

Anderson, J. and Spanglar, G. "Radiometric Dating: Is the 'Decay Constant' Constant?" *Pensee*, Fall 1974.

Barnes, Thomas. *Origin and Destiny of Earth's Magnetic Field.* (Technical Monograph), No. 4, 1973.

Bibby, Geoffrey. "The Big Stones." *Harper's Magazine*, Spring 1973.

Carriveau, G. and Han, M. "Thermoluminescent Dating and the Monsters of Acambaro." *American Antiquity*, 41, 1976.

Driscoll, Evelyn. "Dating of Moon Samples: Pitfalls & Paradoxes." *Science News*, Jan. 1, 1972.

Evans, J.S. "Redating Prehistory in Europe." *Archaeology*, March 1977.

Faul, Henry. "Doubts of the Paleozoic Time Scale." *American Geophysical Union Transactions*, May 1959.

Schiller, R. "When Did Civilization Begin?" *Reader's Digest*, May 1975.

Slusher, Harold. *Critique of Radiometric Dating.* (Technical Monograph), No. 2, 1976.

Velikovsky, I. *Ages in Chaos.* Garden City, N.Y., Doubleday, 1952.

———."Are the Moon's Scars only 3,000 Years old?" *The New York Times*, July 21, 1969.

———."Astronomy & Chronology." *Pensee*, 3, 1973.

———."Pitfalls of Radiocarbon Dating." *Pensee*, Spring, 1973.

Wiant, Harry. "How Reliable Is C-14 Dating?" *Creation Research Society Quarterly*, Jan. 1966.

York, Derek. "Lunar Rocks and Velikovsky's Claims." *Pensee*, May, 1972.

Catastrophism/Uniformitarianism

"The Politics of Science and Dr. Velikovsky." *American Behavioral Scientist*, Sept. 1963.

Bright, John. "Has Archaeology Found Evidence of the Flood?" *Biblical Archaeologist*, Dec. 1942.

Burridge, Gaston. "Collisions from Outer Space." *American Mercury*, April 1962.

Burrow, J.W. "The Flood." *Horizon*, Summer 1972.

Cardona, Dwardu. "The Problem of the Frozen Mammoths." *Kronos*, Winter 1976.

Casson, M. "The Fossil Mystery." *Rocks and Minerals*. No. 396, 1972.

Coffin, H. "Research on Classic Joggins Petrified Trees." *Creation Research Society Quarterly's Annual*, 1969.

Dietz, Robert. "Astroblemes." *Scientific American*, Aug. 1961.

Farrand, William. "Frozen Mammoths and Modern Geology." *Science*, 133, 1961.

Francis, Peter. "Fire and Ice." *New Scientist*, July 3, 1975.

Galloway, William. *The Testimony of Science to the Deluge.* London, 1896.

Gilvarry, J.J. "How the Sky Drove Land from the Sea Bottom." *Saturday Review*, Nov. 4, 1961.

Glass, B. and Heezen, B. "Tektites and Geomagnetic Reversals." *Scientific American*, July 1967.

Gould, Stephen. "Is Uniformitarianism Necessary?" *American Journal of Science*, March 1965.

———. "Is Uniformitarianism Useful?" *Journal of Geological Education*, Oct. 1967.

Hapgood, Charles H. "Mystery of the Frozen Mammoths." *Coronet*, Sept. 1960.

———. *Earth's Shifting Crust*, New York, Pantheon Books, 1958.

Hartmann, William. "Cratering in the Solar System." *Scientific American*, Jan. 1977.

Heylmum, Edgar. "Should We Teach Uniformitarianism?" *Journal of Geological Education*, Jan. 1971.

Hibben, Frank C. *The Lost Americans*. New York, Thomas Y. Crowell Company, 1946.

Howorth, Henry. *The Mammoth and the Flood*. London, S. Low, Marston, Searle, & Rivington, 1887.

Jordan, David S. "A Miocene Catastrophe." *Natural History*, Jan.-Feb., 1920.

King, C. "Catastrophism and Evolution." *American Naturalist*, Aug. 1877.

Krynine, P.D. "Uniformitarianism Is a Dangerous Doctrine." *Journal of Paleontology*, 30, 1956.

Larrabee, Eric. "Scientists in Collision: Was Velikovsky Right?" *Harper's Magazine*, Aug. 1963.

————. "The Day the Sun Stood Still." *Harper's Magazine*, Jan. 1950.

LeRiche, P. "Scientific Proofs of a Universal Deluge." *Transactions of the Victoria Institute*, 1929.

Mackie, Euan. "Megalithic Astronomy and Catastrophism." *Pensee*, Winter 1974-5.

Melton, F.A. "Carolina Bays—Meteorite Scars?" *Journal of Geology*, 1933.

Millman, Peter M. "The Space Scars of Earth." *Nature*, July 16, 1971.

"Moon, Earth Impact Similarities Studied." *Aviation Week and Space Technology*, June 17, 1974.

Morris, Henry M. and Whitcomb, John C. *The Genesis Flood*. Grand Rapids, Baker Book House, 1965 [c 1961].

"Natural Earth Satellite Collided 6,000 Years Ago." *Science Newsletter*, Sept. 18, 1965.

Newell, Norman. "Crises in the History Of Life." *Scientific American*, Feb. 1963.

Oberg, James F. "Baptism of Fire." *Astronomy*, Sept. 1975.

Opik, Ernst. "Catastrophic Approach of a Planet." *Irish Astronomical Journal*, Oct. 1972.

Raikes, R. "Physical Evidence for Noah's Flood." *Iraq*, 28, 1966.

Russell, D. and Tucker, W. "Supernovas and The Extinction of the Dinosaurs." *Nature*, 229, 1971.

Sanderson, Ivan. "Riddle of the Frozen Giants." *Saturday Evening Post*, Jan. 1960.

————. "Riddle of the Frozen Mammoths." *Reader's Digest*, April 1960.

Stewart, John. "Disciplines in Collision." *Harper's Magazine*, June, 1951.

Sullivan, Walter. "Soviet 'Crater' Discovery Sign that Earth Heavily Bombarded." *The New York Times*, Oct. 24, 1976.

Valentine, James. "The Present Is the Key to the Present." *Journal of Geological Education*, April, 1966.

Velikovsky, Immanuel. *Earth in Upheaval*. Garden City, N.Y., Doubleday, 1955.

————. *Worlds in Collision*. New York, Macmillan, 1950.

————. "Answer to My Critics." *Harper's Magazine*, June, 1951.

Von Fange, Erich. "Strange Fires on the Earth." *Creation Research Society Quarterly*, Dec. 1975.

West, Ronald. "Paleontology and Uniformitarianism." *Compass*, May 1968.

Wright, Herbert E. and Martin, P.S. *Pleistocene Extinctions, the Search for a Cause*. New Haven, Yale University Press, 1967.

Zvirblis, A.G. "Earth in Upheaval." *Rocks and Minerals*, April 1974.

Paleo-Anthropology
Anomalous Artifacts

Becker, George. "Antiquities from Under Tuolumne Table Mountain, California." *Bulletin of the Geological Society of America,* Feb. 20, 1891.

Bird, Annie Laurie. *Boise, the Peace Valley.* Caldwell, Idaho, The Caxton Printers, Ltd., 1934.

Blake, W.P. "The Pliocene Skull and Stone Implements of Table Mountain." *Journal of Geology,* Oct.-Nov. 1899.

Brewster, David. "Statements Concerning a Nail Found Imbedded in Sandstone from Kingoodie Quarry, North Britain." *Report of the British Association,* 1844.

Calvert, Frank. "On the Probable Existence of Man During Miocene Period." *Journal of the Royal Anthropology Institute,* v. 3, 1874.

"Cosmology—Meteorite Found in Tertiary Coal." *Comptes Rendus,* v. 193, 1887.

"Curious Geological Facts." *American Journal of Science,* 1, 1820.

"Discovery of Iron Instrument Imbedded in Coal Seam in Glasgow." *Proceedings of the Society of Antiquities of Scotland,* 1, 1.

Dubois, William. "On a Quasi-Coin Reported Found in Boring in Illinois." *American Philosophical Society Proceedings,* Dec. 1871.

Fisher, O. "Fossil Deer's Horn Showing Marks of Human Operation." *The Geologist,* July 1861.

"A Fossil Stone Wall." *Scientific American,* Jan. 14, 1886.

Henshaw, H.W. "Archaeologic Discovery in Idaho." *American Anthropologist,* April 1890.

———. "Preliminary Revision of Evidence of Man in Auriferous Gravels." *American Anthropologist,* 1899.

———. "Review of Evidence Relating to Auriferous Gravel Man." *Smithsonian Annual Report,* 1899.

Hughes, T.M. "Man in the Crag." *Geological Magazine*, June 1872.

Jochmans, J.R. "Out-of-Place Metal Objects Found in Sedimentary Rocks." *Ooparchist*, Oct.-Nov. 1976.

————. "Out-of-Place Elements of High Civilization Found in Rock Strata." *Lost Origins* . . . Privately published, 1976.

Malthaner, Hubert. "Not the Salsburg Steel Cube, but an Iron Object from Wolfsegg." *Pursuit*, Oct. 1973.

"Meteorite in Coal." *Harwicke's Science Gossip*, 1887.

Moir, J. Reid. "A Remarkable Object from Beneath the Red Crag." *Man*, April 1929.

————. "A Piece of Carved Chalk from Suffolk." *Man*, Dec.-Feb. 1919.

————. *Antiquity of Man in East Anglia*. Cambridge, England, The University Press, 1927.

"The Nampa Image." *Science Digest*, Sept. 1970.

Nature, Nov. 11, 1886. Artificially-Shaped Tertiary Meteorite.

"A Necklace of Prehistoric God." Morrisonville *Times*. June 11, 1891.

"Peculiar Stone Found." *Daily Register*, April 6, 1897.

"Personal and Scientific News." *American Geologist*, Dec. 1889.

Powell, O.W. "Are These Evidences of Man in Glacial Gravels?" *Popular Science Monthly*, July 1893.

Proceedings of the Boston Society of Natural History, Jan. 1, 1890. Data on the "Nampa Image."

Rawson, A.L. "Ancient Inscription on a Wall at Chatata, Tennessee." *Transactions of the New York Academy of Science*, 11, 1891.

"A Relic of a Bygone Age." *Scientific American*, June 1852.

"Report and Discussion on Artificially Drilled Shark's Teeth in Pliocene." *Journal of the Anthropology Institute*, 2, 1873.

Skertchly, S. "The Occurrence of Stone Mortars in Ancient Pliocene Gravels of California." *Journal of the Anthropological Institute*, May 1888.

Stopes, Marie. "Human Art in the Red Crag." *Geological Magazine*, Feb. 1912.

"Strange Inscription in a Coal Mine." Los Angeles *Times*, Dec. 17, 1869.

Talmage, J.E. "Notes Concerning a Peculiarly Marked Sedimentary Rock." *Journal of Geology*, 4, 1896.

Wallace, Alfred. "The Antiquity of Man in North America." *19th Century*, Nov. 1887.

Winchell, A. *Sparks from a Geologist's Hammer.* Chicago, S.C. Griggs and Company, 1881.

Wright, F.W. *American Geologist*, April 1899; Correspondence regarding the "Nampa Image."

Wright, G.F. "Additional Notes on the Nampa Image." *Proceedings of the Boston Society of Natural History*, 25, 1891.

———. "The Idaho Find." *American Antiquity*, 11, 1889.

———. "An Archaeological Discovery in Idaho." *Scribners*, Jan. 1890.

———. *Man and the Glacial Period.* Akron, The Werner Company, 1892.

———. "Table Mountain Archaeology." *Nation*, May 21, 1891.

Man and Dinosaur

"Bushmen's Paintings Baffling to Scientists." *Evening News*, Jan. 1, 1970.

Dipeso, Charles. "Clay Figurines of Acambaro, Mexico." *American Antiquity*, 4, 1953.

———. "The Clay Monsters of Acambaro." *Archaeology*, Summer 1953.

———. "Archaeological Quandry." *Fortnight*, Nov. 12, 1951.

Drum, Ryan. "The Cabrera Rocks." *INFO Journal*, May 1976.

Harmer, Lowell. "Mexican Finds Give Hint of Lost World." Los Angeles *Times*, March 25, 1951.

"The Hava Supai Canyon 'Dinosaur.' " *Pursuit*, Jan. 1975.

"The Julsrud Ceramic Collection in Acambaro." *Pursuit*, April, 1973.

"The Serpent Bird of the Mayas." *Science Digest*, Nov. 1968.

Springstead, W. "Herodotus, the Bible, and Flying Serpents." *Bible Science Newsletter*, May 1971.

Tompkins, Peter. "Did Men Fight Dinosaurs?" *New Age Journal*, Dec. 1974.

Verrill, A.H. *Strange Prehistoric Animals and Their Stories.* Boston, L.C. Page, 1948.

Willis, Ronald. "The Acambaro Figurines." *INFO Journal* 2, No. 2

Eolithology

Abbott, W.L. "Worked Flints from Cromer Bed." *Natural Science*, Feb. 1897.

Bennett, F.J. "Home-Made Natural Eoliths." *Geological Magazine*, Jan. 1913.

Burkitt, M.C. "Eoliths: A Test Specimen." *Man*, Aug. 1932.

Burthoud, E.L. "On Prehistoric Human Art from Wyoming and Colorado." *Proceedings of the Academy of Natural Science of Philadelphia*, June, 1872.

Charlesworth, J.K. *The Quarternary Era* . . . "Early Man." London, 1957.

Claypole, E.W. "Human Relics in the Drift of Ohio." *American Geologist*, Nov. 1896.

Cole, G. "Miocene Man in Burma." *Natural Science*, Oct. 1895.

"Did Man Exist in the Tertiary Age?" *American Naturalist*, March 1871.

"Earlier Americans?" *Newsweek*, Nov. 26, 1973.

Jones, Rupert. "Miocene Man in India." *Natural Science*, Nov. 1894.

Lubbock, John. "Existence of Man in the Miocene." *Nature*, March 27, 1873.

MacCurdy, G.G. "The Eolithic Problem." *American Anthropologist*, v. 7, No. 3, 1905.

MacDonald, William. "How Old Is American Man?" *National Observer*, May 31, 1975.

Oldham, R.D. "Alleged Miocene Man at Burma." *Natural Science*, Sept. 1895.

Moir, J. Reid. *The Antiquity of Man in East Anglia*. Cambridge, England, The University Press, 1927.

———. "Discoveries of Human-Fashioned Flints in Red Crag." *Proceedings of the Prehistory Society*, 3, 1921.

———. "The Problem of Miocene Man." *Discovery*, Oct. 1925.

———. "Tertiary Man in England." *Natural History*, Nov.-Dec. 1924.

Noetling, Fritz. "Chipped Flints in the Upper Miocene of Burma." *Records of the Geological Survey of India*, May 1894.

Ephemeral Fossils   (Footprints)

Adams, W.A. "Footprints and Artificial Impressions on Rocks." *American Journal of Science*, 44, 1843.

"The Age of the Nicaragua Footprints." *American Antiquarian*, 11, 1889.

Allen, E.A. "Footmarks in Kentucky." *American Antiquarian*, 7, 1885.

Barnaby, W.O. "250,000-Year-Old Human Footprint." *Nature*, April 17, 1975.

Bird, Roland T. "Thunder in His Footsteps." *Natural History*, May 1930.

Briton, D.G. "On Ancient Human Footprints in Nicaragua." *Journal of the Victoria Institute*, 22, 1889.

Bryan, Alan. "New Light on Ancient Nicaraguan Footprints." *Archaeology*, June 1973.

Burdick, Clifford. "The Antelope Springs Fossil Beds in Utah." *Bible Science Newsletter*, Aug.-Sept. 1969.

Conte, Joseph. "Carson Footprints." *Nature*, May 31, 1883.

Cook, Melvin. "William Meister Discovery of Human Footprint with Trilobites in Cambrian of Utah." *Creation Research Society Quarterly*, Dec. 1968.

Cope, E.D. "The Nevada Biped Tracks." *American Naturalist*, 17, 1883.

Crawford, J. "Neolithic Man in Nicaragua." *American Geologist*, Sept. 1891.

Daly, Reginald. "Human Footprints in the Coal Age." *Bible Science Newsletter*, April 1970.

Dougherty, Cecil N. *Valley of the Giants*. Privately published, 1971.

Flint, Earl. "Human Footprints in Eocene." *American Antiquity*, 10, 1888.

"Footprints." *National Museum Report*, 1890.

"Footprints in the Stone." *Green Bay Gazette*, May 31, 1897.

"Fossil Footprints." *American Anthropologist*, March 1896.

Haberland, W. and Grebe, W.H. "Prehistoric Footprints from El Salvador." *American Antiquity*, 3, 1957.

Hartman, Emerson. *Giant of the Sierras*. Privately published, 1921.

"Human Footprints—600 Million Years Old?" *Daily Reading Magazine*, July 1975.

"Human Footprints in Stone." *Oklahoma Today*, Spring 1975.

"Human-Like Footprints, 250 Million Years Old." *Berea Alumnus*, Nov. 1938.

"Impressions of Feet in Rock." *American Journal of Science*, 33, 1838.

Ingalls, Albert. "The Carboniferous Mystery." *Scientific American*, Jan. 1940.

Jenkins, R. "Sooner Mystery." Dallas *Times Herald*, Aug. 14, 1975.

Knapp, A. "The Triassic Shoe-Sole." *Doheny Scientific Expedition to Hava Supai Canyon*, Oakland, 1927.

Logan, S.H. "Mysterious Footprints in Rock at Clarksville." *Arkansas Historical Quarterly*, Dec. 1942.

Marsh, O.C. "On Supposed Human Footprints Found in Nevada." *American Journal of Science*, 3, 26.

McKennon, C.H. "But If Not, Then What?" Tulsa *Sun World*, May 25, 1969.

Meister, William J. "Discovery of Trilobites in Shod Human Footprint." *Creation Research Society Quarterly*, Dec. 1968.

Mountain, Edgar. "Footprints in Calcareous Sandstone at Nahoon Point." *South African Journal of Science*, April 1966.

"Pre-Adamite Tracks." *American Antiquarian*, 7, 1885.

"Prehistoric Giant." *Nature*, April 19, 1883.

Rusch, Wilbert. "Human Footprints in Rock." *Creation Research Society Quarterly*, March 1971.

"Report on Texas Tracks" and "Footprints in Oklahoma." *Bible Science Newsletter*, Aug.-Sept. 1969.

Stock, Chester. "Mylodon and its Relation on Problem of Supposed Human Prints Occurring Near Carson, Nevada." *Bulletin of the Geological Society of America*, March 31, 1917.

"Stone Age Footprints, Pits Unearthed at Olduvai Gorge."
National Geographic News Bulletin, 1972.

Tolbert, Frank. "Geologists Study Paluxy 'Man Tracks.'"
Dallas *Morning News*, Aug. 14, 1975.

———. "Tracks of Man-Like Giant Under Waterfall." Ibid.,
Jan. 6, 1973.

Hominid Remains

Ayres, W.O. "The Ancient Man of Calaveras." *American
Naturalist*, 16, 1882.

Barnes, F.A. "The Case of the Bones in Stone." *Desert*, Feb.
1975.

———. "Mine Operations Uncover Puzzling Remains of Ancient Man." *Times Independent*, June 3, 1971.

"Bones of Contention," *Newsweek*, Feb. 3, 1967.

Burdick, Clifford. "Discovery of Human Skeleton in Cretaceous Rock." *Creation Research Society Quarterly*,
Sept. 1973.

Collyer, Robert. "The Fossil Human Jaw from Suffolk." *Anthropological Review*, April 1867.

Cope, E.D. "Pliocene Man." *American Naturalist*, Jan. 1880.

Eiseley, L.C. *The Immense Journey* . . . . "Man of the Future." New York, Random House, 1957.

———. "The Antiquity of Modern Man." *Scientific American*, July 1948.

"Fossil Man in Mexico." *American Naturalist*, Sept. 1884.

Frair, Wayne. "The Human Skull Composed of Coal." *Creation Research Society Quarterly*, March 1969.

Hrdlickla, Ales. "Skeletal Remains Attributed to Early Man
in North America." *Smithsonian Bulletin*, 33, 1907.

"Human Tooth Found in Montana Coal Bed." The New York
Times, Nov. 8, 1926.

Judd and Glass. *Voices from the Rocks.* London, 1857.

Leakey, Richard. "Skull 1470." *National Geographic,* June 1973.

Osborn, Henry F. "Pliocene Man of Foxhall in East Anglia." *Natural History,* Nov.-Dec. 1921.

Schoolcraft, Henry. "Remarks on Human Footprints Observed in Secondary Limestone of the Mississippi Valley," *American Journal of Science,* 5, 1822.

Smith, Nathan. "Fossil Bones Found in Red Sandstone." *American Journal of Science,* 2, 1820.

Stutzer, Otto. "Remains in Coal Beds." *Geology of Coal,* Chicago, 1940.

Wright, G. "Prehistoric Man on the Pacific Coast." *Atlantic Monthly,* April 1891.

Physical Anthropology

Burr, David. "Further Evidence Concerning Speech in Neanderthal Man." *Man,* March 1976.

Cohen, D. "Neanderthal Man: Is He One of Us?" *Science Digest,* Oct. 1968.

"D-Deficient Neanderthal." *Sciences,* Nov. 1970.

Dahlberg, A. and Carbonell, V. "Dentition of Magdalenian Female from Cap Blanc, France." *Man,* March 1961.

Dewar, Douglas. "Recent Discoveries on the Origin of Man." *Victoria Institute,* 1953.

Holloway, Ralph. "Casts of Fossil Hominid Brains." *Scientific American,* July 1974.

Ivanhoe, Francis. "Was Virchow Right About Neanderthal?" *Nature,* Aug. 8, 1970.

Johanson, Donald. "Ethiopia Yields First 'Family' of Early Man." *National Geographic,* Dec. 1976.

Langford, E. "New Evidence in Anthropology." *American Mercury,* Spring 1974.

"Leakey's New Skull Changes Our Pedigree." *Science News*, Nov. 18, 1972.

Martindale, Robert. "Oklahoman Says 'Oldest Man' Found in 1925." Tulsa *World*, May 21, 1977.

Montagu, Ashley. "Facial Reconstructions." *American Journal of Physical Anthropology*, 6, 1948.

"Neanderthal Man, Victim of Malnutrition?" *Prevention*, Oct. 1971.

'The Old Man." *Science Digest*, Feb. 1975.

Osborn, Henry. "Is the Ape-Man a Myth?" *Human Biology*, 1, No. 1.

Smith, E. "Neanderthal Man Not Our Ancestor." *Scientific American*, Aug. 1928.

Tobias, Phillip. "The Taung Skull Revisited." *Natural History*, Dec. 1974.

Anthropometry (and Microlithology)

Argyll, W. "Prehistoric Giants." *Nature*, April 19, 1883.

Beirle, F.P. *Giant Man Tracks*. Privately published, 1974.

"Bronze Age Tomb Yields 7 Foot Body." *Los Angeles Times*, March 29, 1976.

Dutt, W.A. "Small Flint Implements from Bungay." *Nature*, Dec. 5, 1907.

Kendall, H.G.O. "Pygmy Flints." *Man*, 1907.

MacRitchie, David. "Prehistoric Pygmies in Silesia." *Nature*, June 12, 1902.

Moir, J. Reid and Burchell, F. "Diminutive Flint Implements of Pliocene Age." *Antiquaries Journal*, April 1935.

"Monmouth Remains." *Daily Independent*, April 4, 1874.

Pittard, Eugene. "Mousterian Microlith Industry." *L'Anthropologie*, 19, 1908.

"Pygmies." *Gentleman's Magazine*, Aug. 1837.

"A Pygmy Cemetery." *Nature*, Aug. 14, 1937.

Rep, Ed. "Death Valley—Garden of Eden." *Wild West*, June 1970.

Scoyen, E.T. "Were There Giants in Those Days?" *Arizona Highways*, July 1951.

Snow, C.E. "Two Prehistoric Indian Dwarf Skeletons." *Geological Survey of Alabama*, 1943.

Stapler, W.M. "A Mystery in History." *New Jersey Highlander*, Spring 1973.

"The So-Called Pygmy Graves in Tennessee." *Harper's Magazine*, Dec. 1876.

"Tiny Tools of Juet Island." *Science World*, March 15, 1961.

"Whither the Red Giants?" *Science Digest*, Jan. 1977.

"Wyoming Mystery Mummy." *Rocks and Minerals*, Sept. 1974.

Cultural Anthropology

Absolon, Karl. "World's Earliest Portrait—30,000 Years Old." *Illustrated London News*, Oct. 2, 1937.

Dart, Raymond. "The Birth of Symbology." *African Studies*, 27, 1968.

Evans, Laura. "The Bones of a Science Writer." *The Sciences*, Sept. 1972.

Haddon, A.C. "The Earliest Known Seat of Learning in Europe." *Natural Science*, Jan. 1897.

LeMay, M. "Language Capability of Neanderthal Man." *American Journal of Physical Anthropology*, Jan. 1975.

Leroi-Gourhan, A. "Flowers Found with Shanidar IV: Neanderthal Burial In Iraq." *Science*, Nov. 7, 1975.

Littauer, Mary. "On Upper Paleolithic Engraving." *Current Anthropology*, Sept. 1974.

McMullen, Roy. "The Lascaux Puzzle." *Horizon*, Spring 1969.

Marshack, Alexander. "The Baton of Montgaudier." *Natural History*, March 1970.

———. "Bone Markings Indicate Ice Age Notational System." The New York *Times*, Jan. 20, 1971.

———. "Cognitive Aspects of Upper Paleolithic Engraving." *Current Anthropology*, June-Oct., 1972.

———. "Exploring the Mind of Ice Age Man." *National Geographic*, Jan. 1975.

———. "Implications of Paleolithic Symbolic Evidence for Origin of Language." *American Scientist*, March-April 1976.

———. "Lunar Notation on Upper Paleolithic Remains." *Science*, Nov. 6, 1964.

———. "The Message in the Markings." *Horizon*, Autumn 1977.

———. "Upper Paleolithic Notation & Symbol." *Science*, Nov. 24, 1972.

Trotter, R. "Tracing the Roots of Civilization." *Science Creation Institute Newsletter*, Feb. 19, 1972.

Wernick, Robert. "Danubian Minicivilization Bloomed Before Ancient Egypt and China." *Smithsonian Magazine*, March 1975.

Prehistoric Technology

Beaumont, P. "The Ancient Pigment Mines of South Africa." *South African Journal of Science*, May 1973.

———. "Amazing Antiquity of Mining in South Africa." *Nature*, Oct. 28, 1967.

Boshier, Adrian. "Swaziland: Birthplace of Modern Man." *Science Digest*, March 1973.

——— and Beaumont, P. "Mining in South Africa and the Emergence of Modern Man." *Optima*, 2, 1972.

Coon, Carleton. "The Origin of Races." *Harper's Magazine,* Dec. 1962.

Dart, Raymond. "The Antiquity of Mining in Africa." *South African Journal of Science,* June 1967.

―――. "Multimillennial Prehistory of Orche Mining." *South African Journal of Science,* Aug. 1968.

――― and Beaumont, P. "Evidence of S. African Iron Ore Mining in Middle Stone Age." *Current Anthropology,* 10, 1969.

―――――――. "Swaziland Iron Ore Mining Radiocarbon Dating," *South African Journal of Science,* June 1968.

Evans, Laura. "Stone-Age Science." *The Sciences,* Nov. 1971.

Lahren, L. and Ronnichson, R. "Bone Foreshafts from Clovis Burial in Montana." *Science,* Oct. 11, 1974.

Lathrop, J. H. "Prehistoric Mines of Lake Superior." *American Antiquities,* July-Aug. 1901.

Moir, J. Reid. "The Problem of Paleolithic Pottery." *Man,* June, 1935.

―――. "Upper Paleolithic Pottery from Ipswich and Swanscombe." *Man,* Nov. 1934.

"Stone Age Astronomers." *Moscow News,* Sept. 4, 1965.

Cartography and Exploration

"Africans Visited North America 2,500 Years Ago." Winnipeg *Free Press,* April 10, 1975.

"Before Columbus or Erikson, Maybe Itanno." The New York *Times,* May 4, 1975.

Carter, George. "Mystery of Indian Civilization." *Science Digest,* May 1957.

Chard, C. S. "Pre-Columbian Trade Between North and South America." *Anthropological Society Papers,* No. 1, 1950.

Colligan, D. "Brawl Over 2,000-Year-Old Archaeological Site." *Science Digest*, Jan. 1973.

Dempewolff, R. "Palenque: Mayan City Inspired by Ancient East." *Science Digest*, Aug. 1968.

Dixon, J. E. et. al. "Obsidian and the Origins of Trade." *Scientific American*, March 1968.

Ekholm, G. "Is American Indian Culture Asiatic?" *Natural History*, Oct. 1950.

Farley, Gloria. "The Stones Speak." Oklahoma Today, *Winter* 1975–76.

Fell, Barry. *America B.C.* New York, Quadrangle/New York Times Book Co., 1976.

Ford, Barbara. "The Brazil Stone: Old Mystery Starts New Fight." *Science Digest*, April 1973.

———. "Semites First in America?" Ibid., 1972.

Gordon, Cyrus. "Authenticity of Parahyba Phoenician Text." *Orientalia*, 37, 1968.

———. *Before Columbus*. New York, Crown, 1971.

———. "Tennessee Stone Inscription Proves America Discovered 1,500 Years Before Columbus." *Argosy*, Jan. 1971.

Hapgood, Charles. *Maps of the Ancient Sea Kings*. Philadelphia, Clinton Books, 1966.

Heyerdahl, Thor. *Aku-Aku, the Secret of Easter Island*, Rand McNally, 1958.

———. *The Ra Expeditions*. Garden City, New York, Doubleday, 1971.

"Inscriptions Hint Phoenicians on Coast 2,500 Years Ago." Hartford *Courier*, April 17, 1975.

Irwin, Constance. *Fair Gods and Stone Places*. New York, St. Martin's Press, 1963.

Jefferys, M. "Pre-Columbian Negroes in America." *Scientia*, July–Aug. 1953.

Lal, Cha-Men. "Did Hindu Sailors Get There Before Columbus?" *Asia*, March 11, 1962.

Mack, Bill. "Ancient Roman Settlement in America?" *Argosy*, March 1972.

Majors, R. H. "The Voyage of the Venetian Brothers Zeno." *Massachusetts Historical Society Proceedings*, Oct. 1874.

Mallery, Arlington. *Lost America.* Washington, D.C., Overlook Co., 1951.

"Map of 1,500 B.C. Shows America." *Reuters News Service*, July 17, 1970.

McKern, S. and Thomas. "Odyssey: Peopling of the New World." *Mankind*, April 1970.

McKern, W.C. "Hypothesis for Asiatic Origin of Woodland Pattern." *American Antiquity*, 3, 1937.

Meggers, Betty and Evans. "Transpacific Contact in 3,000 B.C." *Scientific American*, Jan. 1966.

——————. "Transpacific Origin of Mesoamerican Civilization." *American Anthropologist*, 77, 1975.

"Obsidian Dates Sea Trading to 7,500 B.C." The New York *Times*, Jan. 23, 1971.

Rothrock, George. "Early Mapping of Land and Sea." *Natural History*, Feb. 1966.

Samz, Jane. "Who Really Discovered America?" *Science World*, March 6, 1975.

Short, J.T. "Claims to the Discovery of America." *Galaxy*, 20, 1875.

Speck, Gorden. "The Journey of Hwui Shan." *Pacific Search*, Oct. 1970.

Warren, M. *New and Old Discoveries in Antarctica* (transcript). Aug. 26, 19xx.

Wells, C. "Discovery of America by the Chinese." *Gentlemen's Magazine*, 3, 19xx.

Chemistry and Agriculture

Bourjaily, Vance. "The Corn of Coxcatlan." *Horizon*, Spring 1966.

Cook. O.F. "Staircase Farms of the Ancients." *National Geographic*, May 1916.

Easby, Dudley. "Ancient American Goldsmiths." *Natural History*, Oct. 1956.

———. "Early Metallurgy in the New World." *Scientific American*, April 1966.

Evenari, M. and Koller, D. "Ancient Masters of the Desert." *Scientific American*, April 1956.

Flannery, Kent. "Ecology and Early Food Production in Mesopotamia." *Science*, 147, 1965.

Hammond, Philip. "Desert Waterworks of the Ancient Nabataeans." *Natural History*, June–July 1967.

Harris, L. "Biochemical Discovery of Ancient Babylonians." *Nature*, March 10, 1923.

Haury, Emil. "Arizona's Ancient Irrigation Builders." *Natural History*, Sept. 1945.

Hodge, F.W. "Ancient Irrigation in Arizona." *American Journal of Physical Anthropology*, July 1893.

Hough, J. and Tanton, T. "The Old Copper Assembly and Extinct Animals." *American Antiquity*, v. 20, No. 2, 1954.

Humphreys, Cyril. "Ancient Gold-Work." *Harper's Magazine*, July 1892.

Kokatnur, Vaman. "Chemical Warfare in Ancient India." *Journal of Chemical Education*, May 1948.

Maryon, Herbert. "Metallurgy of Gold and Platinum in Pre-Columbian Ecuador." *Man*, Nov.-Dec. 1941.

Micholic, S. "Art Chemistry." *Science Monthly*, Dec. 1946.

Morey, George. "Problems of Ancient Glassware." *Discovery*, Jan.-Dec. 1930.

Parsons, J. and Denevan, William. "Pre-Columbian Ridged Fields." *Scientific American*, July 1967.

Pfeiffer, John. "Man's First Revolution." *Horizon*, Sept. 1962.

Ray, P.R. "Chemistry in Ancient India." *Journal of Chemical Education*, June 1948.

Reiss, Otto. "Why Was This 9-Ton Slab of Glass Made?" *Art and Archaeology Newsletter*, April-May 1966.

Smith, Ernest. "Solders Used by Goldsmiths of Ur." *Discovery*, Jan.-Dec. 1930

Smith, Ray W. "History Revealed in Ancient Glass." *National Geographic*, Sept. 1964.

"Stone Age Had Farmers." *Science News Letter*, May 5, 1951.

Steinbring, Jack. "Old Copper Culture Artifacts." *American Antiquity*, Vol. 31, No. 4, 1966.

Wadsworth, N. "Solving the Mystery of Yucatán's Ancient Canals." *Science Digest*, March 1974.

Wertime, T.A. "Man's First Encounters with Metallurgy." *Science*, 146, 1964.

Woodbury, R. "Hohokam Canals at Pueblo Grande, Arizona." *American Antiquity*, 26, 1960.

Wulff, H.E. "The Qanats of Iran." *Scientific American*, April 1968.

Art and Architecture

"Ancient City Planning on the Yucatán Peninsula." *Science News*, June 2, 1973.

Andrews, E.W. "Dzibilchaltun: Lost City of the Maya." *National Geographic*, Jan. 1959.

Bennett, Wendell. "A Reappraisal of Peruvian Arts." *Archaeological Memoir*, 4, 1948.

Bibliography

Cable, Mary. "Who Built Zimbabwe?" *Horizon*, Spring 1976.

Cohen, D. "Mystery of the Nazca Lines." *Science Digest*, May 1970.

Cummings, Byron. "Cuicuilco and Mexican Archaic Culture." *Bulletin of the University of Arizona*, Nov. 15, 1933.

Gibson, M. "Nippur—New Perspectives." *Archaeology*, Jan. 1977.

Hambly, Wilfrid. "A Walt Disney in Ancient Egypt." *Science Monthly*, Oct. 1954.

Henderson, R. "Giant Desert Figures Restored." *Desert Magazine*, Nov. 1957.

Keatings, R. and Day, K. "Chan Chan." *Archaeology*, Oct. 1974.

Kosok, P. and Reiche, M. "Ancient Drawings on the Desert of Peru." *Archaeology*, Winter 1949.

"The Lost City of Pajaten." *Horizon*, Autumn 1967.

Lucas, A. "Were the Giza Pyramids Painted?" *Antiquity*, 12, 1938.

Marcus, J. "Territorial Organization of the Classic Maya." *Science*, June 1, 1973.

McIntyre, Loren. "Mystery of the Ancient Nazca Lines." *National Geographic*, May 1975.

Newberry, Percy. "Cinematographic Touch in Ancient Egypt: Wall Paintings Suggesting Moving Pictures." *Illustrated London News*, Jan. 12, 1929.

Perrot, Jean. "Cave-Dwelling Carvers of 5,000 Years Ago." *Horizon*, Jan. 1962.

Pfeiffer, John. "America's First City." *Horizon*, Spring 1974.

Reiche, Maria. *Mystery on the Desert*. Stuttgart-Vaihingen, 1976.

Verrill, A.H. "Oldest City in the World." *Travel*, Sept. 1929.

Whitcomb, Ben. "Lost Pyramids of Rock Lake." *Skin Diver*, Jan. 1970.

Biology and Medicine

Biggs, R. "Medicine in Ancient Mesopotamia." *History of Science*, 8, 1969.

Bokonyi, S. "Development of Early Stock Rearing in the Near East." *Nature*, Nov. 4, 1976.

Brewer, Sam P. "Inca Tools Used in Successful Brain Operation." The New York *Times*, Oct. 4, 1953.

Brothnvell, D.R. and Christensen, M. "A Possible Case of Amputation, Circa 2,000 B.C." *Man*, Dec. 1963.

Davies, R.W. "Medicine in Ancient Rome." *History Today*, 21, 1971.

"Etruscan Artificial Teeth." *Nature*, April 16, 1885.

Freeman, Leonard. "Surgery of the Ancient Inhabitants of America." *Art and Archaeology*, Aug. 1924.

Frisch, Bruce. "The 'Diseased' Statues of Ancient Mexico." *Science Digest*, Feb. 1968.

Gordon, C.A. "Notes on Medical Knowledge in Ancient India." *Transactions of the Victoria Institute*, 25, 1891–2.

Holder, P. and Stewart, T.D. "Complete Find of Filed Teeth from Cahokia Mound, Illinois." *Journal of the Washington Academy of Science*, v. 48, 11.

Hooton, E.A. "Oral Surgery in Egypt During the Old Empire." *Harvard African Studies*, 1, 1917.

Hrdlicka, Ales. "Trepanation among Prehistoric People." *Ciba Symposia*, 1, 1939.

Hsieh, E. "A Review of Ancient Chinese Anatomy." *Anatomical Record*, 20, 1921.

Jastrow, M. "Babylonian-Assyrian Medicine." *Annals of Medical History*, 1, 1917.

Leek, F.F. "Did a Dental Professor Exist in Ancient Egypt?" *Dental Delineator*, Spring 1969.

Linne, Sigvad. "Technical Secrets of American Indians." *Journal of the Royal Anthropology Institute*, July-Dec. 1957.

Lint, J.G. "Treatment of Abdominal Wounds in Ancient Egypt." *Annals of Medical History*, 9, 1927.

MacCurdy, G. "Surgery among Ancient Peruvians." *Art and Archaeology*, Dec. 1910.

Marley, Faye. "Pre-Columbian Medicine Men—Relics Show Their Merits." *Science Newsletter*, Nov. 12, 1966.

McKern, S. and Thomas McKern. "Brain Surgery in the Stone Age." *Science Digest*, Feb. 1970.

——————. "Prehistoric Disease and Primitive Medicine." *Mankind*, v.2, No. 1, 1970.

Mouth-to-Mouth resuscitation, ancient methods of, described. *Science Journal*, June 1967.

Perino, G. "Additional Discoveries of Filed Teeth in Cahokia Area." *American Antiquity*, v. 32, No. 4, 1967.

Protsch, R. and Berger, R. "Earliest Dates for Domesticated Animals." *Science*, 179, 1973.

Senn, N. "Pompeian Surgery and Surgical Instruments." *Medical News*, Dec. 28, 1895.

Shapiro, H.L. "Primitive Surgery." *Natural History*, May-June, 1921.

Smith, G.E. "The Most Ancient Splints." *British Medical Journal*, 1, 1908.

Stewart, T.D. "Significance of Osteitis in Ancient Peruvian Trephining." *Medical History Bulletin*, July-Aug. 1956.

Sweet, A. et al. "Pre-Hispanic Dentistry." *Dental Radiography and Photography* 3, 1.

Thomas, R.E. "Re-invention in the World of Surgery." *Scientific American*, Oct. 20, 1900.

"Trepanning Among the Incas." *Anthropology Review*, 1, 1871-2.

Vieth, I. "Surgical Achievements of Ancient India." *Surgery*, 49, 1961.

Wallis, C.E. "Ancient & Modern Dentistry." *Science Progress*, 1911.

Wilkinson, Richard. "Techniques of Ancient Skull Surgery."
Natural History, Oct. 1975.
Wilson, J.A. "Medicine in Ancient Egypt." *Medical History
Bulletin*, 36, 1962.

Astronomy and Mathematics

Alexander, George. "The Search for Ancient Astronomers."
Los Angeles *Times*, June 12, 1975.
"Ancient Records in Modern Chinese Science." *New Scientist*, Jan. 29, 1976.
Anderson, F. "Arithmetic in Maya Numerals." *American
Antiquity*, 36, 1971.
Angell, I.O. "Stone Circles: Megalithic Math or Neolithic
Nonsense?" *Mathematics Gazette*, Oct. 1976.
Aveni, A.F. et al. "Chichén Itzá's Caracol Tower: Ancient
Astronomical Observatory." *Science*, June 6, 1975.
Baity, Elizabeth. "Archeoastronomy and Ethnoastronomy
So Far." *Current Anthropology*, 1973.
Beach, A.D. "Stonehenge I and Lunar Dynamics." *Nature*,
Jan. 6, 1977.
Brincherhoff, R. "Astronomically Oriented Markings on
Stonehenge." *Nature*, Oct. 7, 1976.
"A Calendar Mosaic from 1,000 B.C." *Science News*, June
30, 1973.
Chatley, Herbert. "Egyptian Astronomy." *Journal of
Archaeology*, 26, 1940.
Colligan, D. "Mayan Astrology: Science of a Super Civilization." *Science Digest*, Feb. 1974.
Colton, R. and Martin, R.L. "Eclipse Cycles and Eclipses At
Stonehenge." *Nature*, Feb. 4, 1976.
——————. "Eclipse Predictions at Stonehenge." Ibid.,
221, 1969.
"A Compass in Central America." *Sciences*, Oct. 9, 1970.

Cowan, T. "Megalithic Rings: Their Design Construction." *Science*, April 17, 1970.

Dow, J.W. "Astronomical Orientation at Teotihuacan." *American Antiquity*, 1967.

Freeman, P.R. "Bayesian Analysis of Megalithic Yard." *Journal of the Royal Statistical Society*, 1976.

Gribbin, John. "Chinese Cosmology." *Astronomy*, Jan. 1977.

Hammerton, M. "The Megalithic Fathom." *Antiquity*, 45, 1971.

Hardman, Clark. "Primitive Solar Observatory at Crystal R. and Its Implications." *Florida Anthropologist*, Dec. 1971.

Hawkins, Gerald. "Astro-Archaeology." *Vistas in Astronomy*, 10, 1968.

———. "Callanish, a Scottish Stonehenge." *Science*, Jan. 1965.

Heggie, D.C. "Megalithic Lunar Observatories: An Astronomer's View." *Antiquity*, 1972.

Henseling, R. "The Scope and Antiquity of Mayan Astronomy." *Research and Progress*, 4, 1938.

Hicks, Robert. "Astronomy in the Ancient Americas." *Sky and Telescope*, June 1976.

———. "Celestial Clues to Egyptian Riddles." *Natural History*, April 1974.

———. "Stonehenge: A Neolithic Computer." *Nature*, June 27, 1964.

———. "Sun, Moon, Men and Stars." *American Scientist*, 53, 1965.

"A Hindu Decimal Ruler of the Third Millennium." *Isis*, v. 25, No. 70, 1936.

Hoyle, Fred. "Stonehenge—an Eclipse Predictor." *Nature*, July 30, 1966.

Jochmans, J.R. *Measuring the Macrocosm*. Privately published, 1975.

Jones, T.B. "Bookeeping in Ancient Sumer." *Archaeology*, Spring 1956.

Kahn, C.H. "On Early Greek Astronomy." *J. Hellenic Studies*, v. 60, 1970.

Krupp, E.C. "Stonehenge: The New Astronomy." *Griffith Observer*, April 1976.

Kuttner, Robert. "Forgotten Sciences." *American Mercury*, Fall 1973.

Libassi, Paul. "Observatories Without Telescopes." *The Sciences*, April 1976.

Linsley, R. & Aveni, A. "Monte Alban: Possible Astronomical Orientation." *American Antiquity* 37, 4, 1972.

Lockyer, Norman. "Some Points on the Early History of Astronomy." *Nature*, May 7, 1871.

"Medicine Wheel Alignments." *Sky and Telescope*, Aug. 1974.

Newham, C.A. "Stonehenge—A Neolithic Observatory." *Nature*, July 30, 1966.

Nordenskiold, E. "Ancient Peruvian System of Weights." *Man*, Dec. 1930.

Patrick, J. "Midwinter Sunrise at Newgrange." *Nature*, June 7, 1974.

Price, Derek J. DeSolla. "An Ancient Greek Computer." *Scientific American*, June, 1959.

———. "The Tower of the Winds." *Natural History* , April 1967.

———. "Unworldly Mechanics." *Natural History*, March 1962.

Rensberger, Boyce. "Rock Art Shows a Supernova." The New York *Times*, Sept. 10, 1976.

Ricketson. O.G. "Astronomical Observatories in the Maya Area." *Geographical Revue*, 18, 1928.

Rodriguez, Luis. "Ancient Astronomy in Mexico and Central America." *Cosmic Perspective*, Jan.-Feb. 1975.

Roger, Ian. "Megalithic Mathematics." *The Listener*, Nov. 27, 1969.

Saville, M.H. "Astronomical Observatories in Ancient Mexico." *Indian Notes*, 6, 1929.

Schove, D.J. "Astro-archaeology Symposium." *Journal of the British Astronomical Association*, 80, 1970.

Smiley, Charles. "The Antiquity and Precision of Mayan Astro." *Journal of the Royal Astronomical Society of Canada*, 1960.

Spinden, H. "Ancient Mayan Astronomy." *Scientific American*, Jan. 1928.

Somerville, B.T. "Megaliths and Astronomy." *Man*, 22, 1922.

———. "Astronomical Indications in Megalithic Callanish." *Journal of the British Astronomical Association*, 23, 1912.

Speicher, John. "The Geometry of Megalithic Man." *Math Gazette*, 45, 1961.

———. "Larger Units of Length of Megalithic Man." *Journal of the Royal Statistical Society*, 127, 4.

———. "Megalithic Astronomy: Indications in Stone." *Vistas in Astronomy*, 7, 1966.

———. "The Megalithic Unit of Length." *Journal of the Royal Statistical Society*, v. 125, 2.

———. "Megaliths and Mathematics." *Antiquity*, June 1966.

———. "Megaliths—Stone Age Mystery." *Science World*, Nov. 4, 1966.

———. "Observing the Moon in Megalithic Times." *Journal of the British Association*, 80, 1970.

———. "Solar Observatories of Megalithic Man." *Journal of the British Astronomical Society*, 64, 1954.

Trotter, Robert. "Stonehenge as an Astronomical Instrument." *Antiquity*, 1, 1927.

Wittry, Warren. "An American Woodhenge." *Cranbrook Institute of Science Newsletter*, 33, 1964.

Engineering Sciences (and General Technology)

Bandelier, A.F. "The Ruins of Tiahuanaco." *American Antiquarian Society. Proceedings*, 21, 1911.

Banks, Edgar. "The Colossus of Rhodes." *Art and Archaeology*, May 1917.

Bromehead, C.N. "Ancient Mining Processes." *Antiquity*, Sept. 1942.

Carter, George, "Secrets of Ancient Engineering." *Science Digest*, Sept. 1960.

Davies, G. and Glanville, S. "Life in Ancient Egypt: Astonishing Skill in Arts & Crafts." *Illustrated London News*, April 12, 1930.

Dubberstein, Waldo. "Babylonians Merit Honor as Original 'Fathers of Science.' " *Science Newsletter*, Sept. 4, 1937.

Dunham, Dow. "Building an Egyptian Pyramid." *Archaeology*, Autumn 1956.

Ekholm, Gordon F. "Wheeled Toys in Mexico." *American Antiquity*, 4, 1946.

Fisher, A. "Machu Picchu—Lost City of the Incas." *Science World*, May 19, 1967.

Goodfield, June. "The Tunnel of Eupalinus." *Scientific American*, 6, 1964.

Grieder, Terence. "Rotary Tools in Ancient Peru." *Archaeology*, July 1975.

Harrison, James O. "Riddle of Costa Rica's Stones Spheres." *Science Digest*, June 1967.

Holz, Peter. "Riddle of Zimbabwe." *Natural History*, May 1956.

Kuh, Katharine. "The Mystifying Maya." *Saturday Review*, June 28, 1969.

Lewis, T.H. "Copper Mines Worked by Mound Builders." *American Antiquity*, 11, 1889.

Lister, Robert. "Additional Evidence of Wheeled Toys in Mexico." *American Antiquity*, 3, 1947.

Lothrop, Eleanor. "Mystery of the Prehistoric Stone Balls." *Natural History*, Sept. 1955.

MacCurdy, George. "Obsidian Razor of the Aztec." *American Anthropologist*, 2, 1900.

Magoffin, R. "Ancients Had 'New' Ideas." *Science Newsletter*, Jan. 21, 1928.

Moseley, Michael. "Secrets of Peru's Ancient Walls." *Natural History*, Jan. 1975.

Newberry, J.S. "Ancient Mining in North America." *American Antiquity*, 11, 1889.

"Prehistoric Masonry Resembling Chinese Wall." *Black Hills Weekly*, South Dakota, Jan. 1. 1896.

Price, D. DeSolla. "Gears from the Greeks." *Transactions of the American Philosophical Society*, Nov. 1974.

Reiss, Otto. "Euripides in the Andes Mountains?" *Arts and Archaeology*, Jan.-March 1967.

Rosenblohm, William. "Geological and Gemologic Thought in Early Classical Times." *Lapidary Journal*, Feb. 1957.

Safer, John. "Los Bolas Grandes: An Archaeological Enigma." *Oceans*, July 1975.

Sassoon, G. and Dale, R. "Deus Est Machina?" *New Science*, April 1976.

Saville, M. "Ancient Maya Causeway of Yucatán." *Antiquity*, March 1935.

Shippee, R. "Air Adventures in Peru." *National Geographic*, Jan. 1933.

Silverberg, Robert. *Wonders of Ancient Chinese Science.* New York, Hawthorn Books, 1969.

Slawson, N.H. "New Light on Ancient Technical Progress." *Scientific American*, July 1935.

Soye, C. Var. "Emperor Caligula Insisted on Quality Valves." *Chemical Engineering*, July 27, 1959.

"Utah Mystery: Prehistoric Mining?" *Coal Age*, Feb. 1954.

Von Hagen, Victor. "Highway of the Inca." *Archaeology*, Summer 1952.

Walters, R.C. "The Groma: Surveying Instrument of the Ancients." *Discovery*, Sept. 1925.

Watkins, J.E. "The Transportation and Lifting of Heavy Bodies by the Ancients." *Smithsonian Annual Report*, 1898.

"Weapons and Politics of the Ancient Hindus." *Nature*, Oct. 21, 1830.

Weigand, Phil. "Mines and Mining Techniques of the Chalchihuites Culture." *American Antiquity*, Jan. 1968.

Aeronautics

"Atlantis and the Searchers." *Newsweek*, July 31, 1967.

"Carthaginian Lenses." *Nature*, Sept. 13, 1930.

"Ancient Aztecs Flew Gliders in Mexico, Archaeologist Reveals." The New York *Times*, July 23, 1934.

Frenchman, M. "2,000-Year-Old Model Glider with Futuristic Look . . ." *The Times*, London, England, May 18, 1972.

"Imhotep's Glider?" *INFO Journal*, Spring 1973.

Jockmans, J.R. *Legacy of Methuselah* . . . "Monoplanes and Megatons." Privately published, 1975.

Laufer, Berthold. *The Prehistory of Aviation*. Chicago, Field Museum of Natural History. 1928.

Mehta, C.N. *The Flight of Hanuman to Lanka via Sunda Islands*. Nadiad, 1941.

"Nazca Balloonists?" *Time*, Dec. 15, 1975.

Salzberg, Ruth. "Solving the Mystery of the Nazca Lines."
Saga, May 1976.
Sanderson, Ivan. "This 'Airplane' Is More Than 1,000 Years
Old!" *Argosy*, Nov. 1969.
Shah, Ikbal. "Aerial Warfare in Ancient India." *Discovery*,
June 1920.

Physics

Brewster, David. "On a Rock Crystal Lens and Glass in
Nineveh." *American Journal of Science*, 15, 1853.
Carlson, John B. "Did Olmecs Possess Magnetic Lodestone
Compass?" *International Congress Amer.*, Sept. 1974.
———. "Lodestone Compass: Chinese or Olmec Primacy."
Science, Sept. 5, 1975.
"An Electric Battery of 2,000 Years Ago." *Discovery*, March
1939.
Foss, C. "A Bullet of Tissaphernes." *Journal of Hellenic
Studies*, 65, 1975.
Fuson, Robert. "The Orientation of Mayan Ceremonial Cen-
ters." *Association of American Geographers. Annals*,
Sept. 1969.
Morant, G.M. "A Morphological Comparison of Two Crystal
Skulls." *Man*, July 1936.
Salm, Walter G. "Babylon Battery." *Popular Electronics*,
July 1964.
Schwalb, Harry. "Electric Batteries of 2,000 Years Ago."
Science Digest, April 1957.
Schweig, Bruno. "Mirrors." *Antiquity*, Sept. 1941.
Shankland, R. "The Development of Architectural Acous-
tics." *American Scientist*, March-April, 1972.
Sutton-Vane, S. "Eyeglasses—Ancient and Modern."
Science Digest, Sept. 1958.
Taylor, H.L. "Origin and Development of Lenses in Ancient

Times." *British Journal of Physiological Optics*, July 1930.

Atlantology

Babcock, William. "Atlantis and Antilla." *Geographical Review*, May 1917.
Balch, Edwin S. "Atlantis or Minoan Crete?" *Geographical Review*, May 1947.
Charlier, Roger. "Perennial Atlantis." *Sea Frontiers*, Jan.-Feb. and March-April, 1972.
Donnelly, Ignatius. *Atlantis: The Antediluvian World*. First printed New York, Harper; 1882; (a modern rev. ed.) New York, Gramercy, 1949.
———. *Ragnarok*. Blauvelt, New York, Rudolf Steiner Publications, 1974.
Gillmer, Thomas. "Ships of Atlantis." *Sea Frontiers*, Nov.-Dec. 1975.
Gonzalez, A. "Lost Island of Atlantis?" *Science Digest*, May 1972.
Greenburg, Lewis. "Atlantis." *Pensee*, Winter 1973-74.
Harrison, W. "Atlantis Undiscovered—Bimini, Bahamas. *Nature*, April 2, 1971.
Hopkins, Albert. "Legendary Islands of the North Atlantic." *Scientific American*, July-Oct. 1921.
Isaacson, I. "Preliminary Remarks about Thera and Atlantis." *Kronos*, Summer 1975.
Jones, Waldo. "Fantasy or Geological Facts?" *Rocks and Minerals*, Jan. 1965.
Kernback, Wilfried. "Atlantis Was in Germany." *American Mercury*, Summer 1972.
Lear, John. "The Volcano that Shaped the Western World." *Saturday Review*, Nov. 5, 1966.

Marx, Robert and Rebikoff, D. "Atlantis at Last?" *Argosy*, Dec. 1969.

——————. "Atlantis: The Legend Becoming Fact." *Argosy*, Nov. 1971.

Matthew, William. "Plato's Atlantis in Palaeography." *Proceedings of the National Academy of Science*, 6, 1920.

Mavor, James W. "Journey to Atlantis." *World of Antiquities*, 1, 1970.

Merrill, E.D. "Scuttling Atlantis and Mu." *American Scholar*, 3, 1936.

Mitchell-Hedges, F.A. "Atlantis Not a Myth but Cradle of American Races." *New York American*, March 10, 1935.

Reynolds, Barbara. "Atlantis 'Found' in Our Own Backyard." Chicago *Tribune*, Oct. 10, 1974.

Roberts, Cokie and Steven. "Atlantis Recaptured." The New York *Times*, Sept. 5, 1976.

Rude, Gilbert. "A Survey of Atlantis." *U.S. Naval Institute Proceedings*, Aug. 1940.

Scharff, R.F. "Some Remarks on the Atlantis Problem." *Proceedings of the Royal Irish Academy*, 24, 1902.

Schuchert, C. "Atlantis and the North Atlantic Ocean Bottom." *Proceedings of the National Academy of Science*, Feb. 1917.

Scranton, Robert. "Lost Atlantis Found Again?" *Archaeology*, Autumn 1949.

Spence, Lewis. *Atlantis Discovered*. New York, Causeway, 1974.

Termier, Pierre. "Atlantis." *Smithsonian Annual Report*, 1915.

Thompson, Edward. "Atlantis Not a Myth." *Popular Science Monthly*, 1879.

ARK-aeology

"A Futile Quest: Search for Noah's Ark." *Biblical Archaeological Revue*, v. 2, No. 2, 1976.

Gaskill, Gordon. "Mystery of Noah's Ark." *Reader's Digest*, Sept. 1975.

Gianone, R. "A Comparison of the Ark with Modern Ships." *Creation Research Society Quarterly*, June 1975.

Jones, Arthur. "How Many Animals in the Ark?" *Creation Research Society Quarterly*, Sept. 1973.

Morris, Henry. "The Ark of Noah." *Creation Research Society Quarterly*, Sept. 1971.

Morris, John. "The Search for Noah's Ark." *Bible and Spade*, Summer 1974.

"Noah's Ark?" *Life*, Sept. 5, 1960.

Odom, Leo. "Has Noah's Ark Been Found?" *Israelite*, April-June 1977.

Pendelton, Paul. "Is It Really Noah's Ark?" *Science World*, Jan. 29, 1973.

Portune, John E. "How Did the Ark Hold 'All Those Animals'?" *Tomorrow's World*, May 1971.

Waltz, E.L. "Diggings from Turkey: Ararat 1974" and "Space-Age ARK-aeology." *Bible and Spade*, Autumn 1974.